SAFETY ASPECTS OF
NUCLEAR POWER PLANTS
IN HUMAN INDUCED
EXTERNAL EVENTS:
ASSESSMENT OF STRUCTURES

The following States are Members of the International Atomic Energy Agency:

AFGHANISTAN	GERMANY	PALAU
ALBANIA	GHANA	PANAMA
ALGERIA	GREECE	PAPUA NEW GUINEA
ANGOLA	GUATEMALA	PARAGUAY
ANTIGUA AND BARBUDA	GUYANA	PERU
ARGENTINA	HAITI	PHILIPPINES
ARMENIA	HOLY SEE	POLAND
AUSTRALIA	HONDURAS	PORTUGAL
AUSTRIA	HUNGARY	QATAR
AZERBAIJAN	ICELAND	REPUBLIC OF MOLDOVA
BAHAMAS	INDIA	ROMANIA
BAHRAIN	INDONESIA	RUSSIAN FEDERATION
BANGLADESH	IRAN, ISLAMIC REPUBLIC OF	RWANDA
BARBADOS	IRAQ	SAINT VINCENT AND
BELARUS	IRELAND	THE GRENADINES
BELGIUM	ISRAEL	SAN MARINO
BELIZE	ITALY	SAUDI ARABIA
BENIN	JAMAICA	SENEGAL
BOLIVIA, PLURINATIONAL	JAPAN	SERBIA
STATE OF	JORDAN	SEYCHELLES
BOSNIA AND HERZEGOVINA	KAZAKHSTAN	SIERRA LEONE
BOTSWANA	KENYA	SINGAPORE
BRAZIL	KOREA, REPUBLIC OF	SLOVAKIA
BRUNEI DARUSSALAM	KUWAIT	SLOVENIA
BULGARIA	KYRGYZSTAN	SOUTH AFRICA
BURKINA FASO	LAO PEOPLE'S DEMOCRATIC	SPAIN
BURUNDI	REPUBLIC	SRI LANKA
CAMBODIA	LATVIA	SUDAN
CAMEROON	LEBANON	SWAZILAND
CANADA	LESOTHO	SWEDEN
CENTRAL AFRICAN	LIBERIA	SWITZERLAND
REPUBLIC	LIBYA	SYRIAN ARAB REPUBLIC
CHAD	LIECHTENSTEIN	TAJIKISTAN
CHILE	LITHUANIA	THAILAND
CHINA	LUXEMBOURG	THE FORMER YUGOSLAV
COLOMBIA	MADAGASCAR	REPUBLIC OF MACEDONIA
CONGO	MALAWI	TOGO
COSTA RICA	MALAYSIA	TRINIDAD AND TOBAGO
CÔTE D'IVOIRE	MALI	TUNISIA
CROATIA	MALTA	TURKEY
CUBA	MARSHALL ISLANDS	TURKMENISTAN
CYPRUS	MAURITANIA	UGANDA
CZECH REPUBLIC	MAURITIUS	UKRAINE
DEMOCRATIC REPUBLIC	MEXICO	UNITED ARAB EMIRATES
OF THE CONGO	MONACO	UNITED KINGDOM OF
DENMARK	MONGOLIA	GREAT BRITAIN AND
DJIBOUTI	MONTENEGRO	NORTHERN IRELAND
DOMINICA	MOROCCO	UNITED REPUBLIC
DOMINICAN REPUBLIC	MOZAMBIQUE	OF TANZANIA
ECUADOR	MYANMAR	UNITED STATES OF AMERICA
EGYPT	NAMIBIA	URUGUAY
EL SALVADOR	NEPAL	UZBEKISTAN
ERITREA	NETHERLANDS	VANUATU
ESTONIA	NEW ZEALAND	VENEZUELA, BOLIVARIAN
ETHIOPIA	NICARAGUA	REPUBLIC OF
FIJI	NIGER	VIET NAM
FINLAND	NIGERIA	YEMEN
FRANCE	NORWAY	ZAMBIA
GABON	OMAN	ZIMBABWE
GEORGIA	PAKISTAN	

The Agency's Statute was approved on 23 October 1956 by the Conference on the Statute of the IAEA held at United Nations Headquarters, New York; it entered into force on 29 July 1957. The Headquarters of the Agency are situated in Vienna. Its principal objective is "to accelerate and enlarge the contribution of atomic energy to peace, health and prosperity throughout the world".

SAFETY REPORTS SERIES No. 87

SAFETY ASPECTS OF NUCLEAR POWER PLANTS IN HUMAN INDUCED EXTERNAL EVENTS: ASSESSMENT OF STRUCTURES

INTERNATIONAL ATOMIC ENERGY AGENCY
VIENNA, 2018

COPYRIGHT NOTICE

All IAEA scientific and technical publications are protected by the terms of the Universal Copyright Convention as adopted in 1952 (Berne) and as revised in 1972 (Paris). The copyright has since been extended by the World Intellectual Property Organization (Geneva) to include electronic and virtual intellectual property. Permission to use whole or parts of texts contained in IAEA publications in printed or electronic form must be obtained and is usually subject to royalty agreements. Proposals for non-commercial reproductions and translations are welcomed and considered on a case-by-case basis. Enquiries should be addressed to the IAEA Publishing Section at:

Marketing and Sales Unit, Publishing Section
International Atomic Energy Agency
Vienna International Centre
PO Box 100
1400 Vienna, Austria
fax: +43 1 2600 29302
tel.: +43 1 2600 22417
email: sales.publications@iaea.org
http://www.iaea.org/books

© IAEA, 2018

Printed by the IAEA in Austria
February 2018
STI/PUB/1769

IAEA Library Cataloguing in Publication Data

Names: International Atomic Energy Agency.
Title: Safety aspects of nuclear power plants in human induced external events: assessment of structures / International Atomic Energy Agency.
Description: Vienna : International Atomic Energy Agency, 2018. | Series: IAEA safety reports series, ISSN 1020–6450 ; no. 87 | Includes bibliographical references.
Identifiers: IAEAL 18-01140 | ISBN 978–92–0–101117–6 (paperback : alk. paper)
Subjects: LCSH: Nuclear power plants — Safety measures. | Nuclear power plants — Risk assessment. | Structural analysis (Engineering).
Classification: UDC 621.039.58 | STI/PUB/1769

FOREWORD

Many human actions pose challenges to the safe operation of a nuclear installation, such as a nuclear power plant. These challenges may arise from activities human beings undertake as a part of routine life. The challenges arising from intentional and accidental events need to be evaluated given the current design robustness of the installation and the vulnerability of the location of such events.

This publication is the second of three Safety Reports on the safety assessment of nuclear facilities subjected to extreme human induced external events. These publications address the assessment of nuclear installations subjected to accidental or unintentional human actions. They provide the general framework for approaches to obtaining the overall plant performance with regard to the fundamental safety functions from the performance of individual components. It includes safety assessments, the characterization and quantification of loadings, and appropriate analysis techniques and material properties for capacity assessments. This publication explores established methodologies in the light of recent advances in the understanding of material behaviour under such extreme loading conditions and computational techniques that can incorporate such behaviour in the analytical modelling.

These three Safety Reports were developed using funding from Member States voluntarily contributing to, and participating in, the extrabudgetary programme of the External Events Safety Section (EESS-EBP). Established in 2007, the EESS-EBP has developed technical documents considered a priority for Member States, given the current experience with severe external events globally. The aim of the programme is to provide technical inputs to current and future IAEA safety standards. The EESS-EBP implements these activities by assimilating the latest technical issues and practical methodologies in Member States, and disseminates the information through technical publications, sharing them in the working groups, and by participating in global conferences and forums.

The work of all the contributors to the drafting and review of this publication is greatly appreciated. In particular, the IAEA gratefully acknowledges the contributions of F.O. Henkel (Germany), S. Hostikka (Finland), N. Krutzik (Germany), N. Orbovic (Canada), R. Ricciuti (Canada) and A. Saarenheimo (Finland) to the drafting of this publication, and of A. Blahoianu (Canada) to its review. The IAEA officers responsible for this publication were A. Altinyollar and F. Beltran of the Division of Nuclear Installation Safety.

CONTENTS

1. INTRODUCTION

1.1. BACKGROUND

The process of safety assessment of a nuclear installation needs to be repeated periodically — in whole or in part, as necessary — in order to take into account changed circumstances with respect to those considered for the design.

Following this principle, the IAEA initiated a major effort in 2001 targeted at the development of guidelines for the assessment of the vulnerability to accidental or postulated human induced external events not foreseen in the design basis. Examples of accidental events include explosions caused by pipeline failures, train crashes and hazardous material leaks from tanks. Examples of postulated external events include station blackouts and loss of ultimate heat sink due to unidentified causes.

There is general agreement among experts that current practice for nuclear power plant design, especially against natural hazards, provides a safety margin and robustness that may enable the plant to withstand some scenarios caused by human induced external events not explicitly considered at the design stage without significant radiological consequences. This is believed to be true for nuclear facilities in general, and for nuclear power plants in particular. However, quantification is needed in order to understand, with a high level of confidence, which events can be screened out in a safety evaluation process and which events require a detailed assessment of the actual plant level performance.

The IAEA has published three Safety Guides that deal with the safety of nuclear power plants against human induced events of accidental origin:

(a) External Human Induced Events in Site Evaluation for Nuclear Power Plants, IAEA Safety Standards Series No. NS-G-3.1 [1];
(b) External Events Excluding Earthquakes in the Design of Nuclear Power Plants, IAEA Safety Standards Series No. NS-G-1.5 [2];
(c) Protection against Internal Fires and Explosions in the Design of Nuclear Power Plants, IAEA Safety Standards Series No. NS-G-1.7 [3].

Reference [1] deals with the examination of a region considered for siting a nuclear power plant, in order to identify hazardous phenomena associated with external human induced events that might occur in the region. It also presents preliminary guidelines for deriving the values of relevant parameters for the design basis. Reference [2] is devoted to the design of protection against the effects of external events, excluding earthquakes. Reference [3] provides design concepts for protection against internal fires and explosions in nuclear power plants.

Structural safety is an important aspect of protection against hazards caused by human induced external events. References [1–3] contain high level guidelines on the characterization of loads arising from hazards caused by human induced external events, such as explosions, impacts or fires. However, they provide almost no guidelines on the structural response analysis for the associated loading effects or on the subsequent evaluation of performance. The loading effects arising out of these hazards are, in general, of an extreme type. The structural response analysis is more complex than the analysis for other static and dynamic loadings. In addition, the structural performance for this type of extreme loading may not be able to be assessed directly using the available design codes for nuclear safety related structures.

Consequently, there is a need to develop detailed guidelines for load characterization as well as for structural response analysis and performance evaluation using the current state of the art in related areas. This Safety Report is intended to meet this need and to cover the advancement of technology since the publication of Refs [1–3].

This publication is the second in a series of three reports that provide guidelines to support the quantitative evaluation of the safety of facilities subjected to design basis events (DBEs) and beyond design basis events:

(a) Safety Aspects of Nuclear Power Plants in Human Induced External Events: General Considerations, Safety Reports Series No. 86 [4];
(b) Safety Aspects of Nuclear Power Plants in Human Induced External Events: Assessment of Structures, Safety Reports Series No. 87;
(c) Safety Aspects of Nuclear Power Plants in Human Induced External Events: Margin Assessment, Safety Reports Series No. 88 [5].

Reference [4] provides the general framework and includes a road map on how to design and evaluate protection against human induced external hazards. The report concentrates on an overview of the methodology and on the important conditions in its application to existing and new nuclear power plants. The topics covered include elements of the design/evaluation approach, developed in five phases:

— Phase 1: Event identification;
— Phase 2: Hazard evaluation and load characterization;
— Phase 3: Design and assessment approaches for structures, systems and components (SSCs);
— Phase 4: Plant performance assessment and acceptance criteria;
— Phase 5: Member State response.

The present Safety Report addresses phases 2 and 3 of the general framework. It gives detailed guidelines for the safety assessment of nuclear power plant structures against mechanical impact, explosion and fire hazards caused by human induced external events. The report covers the characterization of loading; the assessment of structural integrity, using both simplified methods and more elaborated methodologies; and the assessment of induced vibration. Acceptance criteria are given for different failure modes: overall stability, overall bending and shear, local failure modes and induced vibrations. Additionally, since many of the human induced external events may result in a fire, the process of analysing the consequences of a fire is also detailed. Approaches to assessing the barrier fire performance and the fire performance of safety related SSCs are given.

Reference [5] addresses phases 1 and 4 of the general framework. The report describes the procedures for calculating the margins of nuclear power plants in human induced external hazards. Both postulated and accidental hazards are considered. The report focuses on plant and systems performance evaluations. A graded approach for margin assessment is provided. The first grade consists of a deterministic procedure in which, for each scenario, the existence of at least one undamaged success pathway to complying with the fundamental safety function is investigated. This procedure can be extended later on to calculate probability measures, such as conditional core damage probability and the conditional probability of spent fuel pool loss of cooling and spent fuel damage, for the given scenario. In the most elaborated stage, probabilistic safety assessment techniques are introduced, giving consideration to the probabilistic aspects of the hazards and of the capacity of SSC fragility. Event-tree and fault-tree models are used to compute the usual probabilistic safety assessment metrics, such as core damage frequency, large early release frequency, and frequency of loss of spent fuel pool cooling and spent fuel damage.

In summary, these three implementation reports (Refs [4, 5] and this Safety Report) provide methodologies that can be used in the evaluation of the capacity of SSCs of nuclear power plants subjected to extreme, human induced external events and in the assessment of the resulting safety margin of the facilities. The reports may be useful to nuclear facility owners, operators and regulators who need an understanding of the safety issues in relation to human induced events. The reports contain a description of internationally accepted methods applied by the engineering community and some examples that may be useful in the evaluation of the need for plant upgrading. Many references are also provided for more detailed guidelines. For technical background, the reports rely on many IAEA safety standards as well as on relevant books.

The three reports share a common thread. Together, they provide an approach to the assessment against extreme, human induced external events that is fully consistent with the methods used for the evaluation of nuclear facilities subjected to extreme natural events, such as earthquakes and floods.

1.2. OBJECTIVE

In order to ensure the basic safety functions of a nuclear power plant in the event of human induced external events, it is necessary that the structures important to reactor and spent fuel pool safety perform their intended functions, specified in terms of structural integrity, leaktightness and serviceability (to maintain the functionality of systems and components housed by the structures). The objective of this report is to provide detailed guidelines for characterizing and quantifying loads (static and dynamic) arising from the impact of human induced hazards on the protective features, for evaluating the structural response analysis against these loadings, and for assessing the structural performance against its intended function.

This Safety Report provides an overall framework and outlines the main sequences for an assessment of the designed performance of the protective features in meeting their intended functions. The report cites many technical references on which it is based and describes the procedures for implementing many state of the art practices; the scope of this report does not allow for the inclusion of details and specifics.

1.3. SCOPE

This Safety Report provides a methodology and detailed guidelines for the assessment of nuclear power plant structures against mechanical impact, explosion, fires and missile hazards caused by accidental[1] human induced external events. Wilful human induced events, such as military action or industrial sabotage, are not within the scope of the report.

The report covers load characterization, structural response analysis and an assessment of protection of nuclear power plants against human induced external hazards. Detailed guidelines are given about the linear/non-linear analysis of structures for impulsive, impactive and thermal loading; modelling of the loads; material modelling; structural modelling; and consideration of boundary conditions. The report also addresses the approach to structural performance

---

[1] The report distinguishes between purely accidental events (for which a probability of occurrence can be computed) and postulated human induced events (which are defined solely, for instance, for design purposes).

assessment and acceptance criteria for these types of loading scenario. The guidelines provided in this Safety Report are applicable to both new and existing nuclear power plants.

1.4. STRUCTURE

This Safety Report has four sections and three appendices. Section 2 deals with load characterization and its quantification arising from the considered hazards. Material models for both the linear and non-linear behaviour of concrete and steel are described in Section 3. Section 4 describes the different types of structural analysis and simulation techniques that are applicable to these loading conditions. The assessment of performance and acceptance criteria for structural evaluation are addressed in Section 5.

Guidelines for derivation of loading functions caused by military and commercial aircraft crashes are presented in Appendix I. Examples of a simplified method for assessing the structural impact, for the derivation of verified loading functions and for typical dynamic response spectra are given in Appendix II. Finally, guidelines for design and assessment of concrete elements against explosion are given in Appendix III.

2. CHARACTERIZATION OF LOADING

2.1. GENERAL CONSIDERATIONS

This report deals with three generic types of hazard, which are consequences of human induced external events: mechanical impact, explosion and fire. These hazards and associated loadings are summarized in Table 1. The following sections discuss the characterization of the loadings resulting from these hazards.

2.2. MECHANICAL IMPACT

2.2.1. Classification of impact loading

Missiles or projectiles causing mechanical impacts are broadly classified into two categories: (i) hard missiles; and (ii) soft missiles. The categorization is based on the missile's deformability with respect to the deformability of the

TABLE 1. SUMMARY OF HAZARDS AND LOADS CAUSED BY HUMAN
INDUCED EXTERNAL EVENTS

Hazard	Mechanical impact		Explosion			Fire	
	Hard	Soft					
Load	Missile impact load	Missile impact load	Blast or pressure load			Thermal load, smoke	
Load type	Dynamic	Dynamic	Equivalent Static	Impulsive Dynamic	Fireball	Pool fire	
Parameters to characterize load	Missile mass, velocity, missile cross-sectional area	Missile mass, velocity, crushing force	Pressure	Pressure transient	Fuel mass	Fuel mass, burning rate	

impacted object. Soft missiles are also referred to as 'deformable missiles' in the relevant literature. It is possible to have a missile categorized in between these two main categories. One explicit criterion for definition of missile types is given in Ref. [6].

2.2.2. Soft missiles

The loading function due to a soft missile impact is dependent on the mass distribution along the length, the crushing force and the velocity of the missile. The loading is impulsive (impulse driven). There are two approaches to characterizing the loading due to the impact of a soft missile: (i) a continuum model using finite element technique (see Section 4.2.3.4); and (ii) an analytical method according to Riera [7]. The loading function due to a soft missile impact according to the Riera method is:

$$F(t) = P_c[x(t)] - m[x(t)]v_m^2(t) \tag{1}$$

There are two terms in Eq. (1). The first term, $P_c[x(t)]$ represents the crushing of the missile. The crushing force depends on the axial stiffness of the missile and can be assessed by computational methods. It should be noted that compressive force is negative in this equation. The second term represents the mass flow or the momentum of the missile impacting the rigid target and is proportional to the mass per unit length of the missile $m[x(t)]$ and to the square of the impact velocity $v_m(t)$ of the section impinging on the target at the considered moment of time. The variable x is the distance to the contact cross-section measured from the nose of the missile.

It is assumed that the object that the missile hits is infinitely rigid and fixed during the impact and that the loaded area does not undergo any local deformations. It is, furthermore, assumed that the longitudinal axis of the flying object coincides with the direction of flight and that, at the moment of impact, it is perpendicular to the target. A typical plot of $F(t)$, as illustrated in Ref. [7] for the impact of a commercial plane, is reproduced in Fig. 1.

Application of the Riera method to the derivation of the loading function for a Phantom F-4 impact and for a Boeing 747 impact is illustrated in Appendix I.

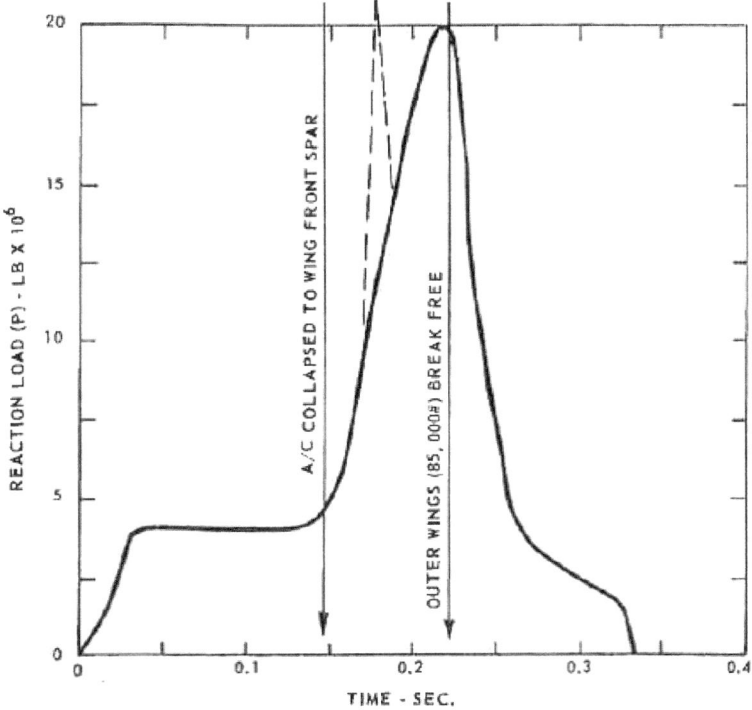

FIG. 1. Force–time function calculated for a Boeing 707-320 [7].

2.2.3. Hard missiles

A hard impact leads to impactive loading (energy driven). In practical applications, structural integrity, in terms of penetration, perforation and scabbing, for hard missile impacts can be assessed by empirical formulas. The formulas, in many cases, provide a practical way of assessing the necessary wall thicknesses. Empirical formulas for assessing penetration, perforation and scabbing are presented in Section 5.3.

2.2.4. Aircraft crashes

Impact loading on a structure or object due to an aircraft crash is a combination of soft, semi-hard and hard missile impacts. The main component in the loading function resulting from a collision of the deformable fuselage can be predicted assuming a soft missile impact. Aircraft engines and landing gear can be classified as semi-hard or hard missiles.

Member States define the stiffness, mass, diameter and velocity of the part of the engine and/or the landing gear to be used in the design and verification. As shown by full scale impact tests on concrete walls, local damage produced by aircraft engines can be predicted by empirical formulas developed to assess local damage caused by the impact of solid (hard) cylindrical missiles, as long as some additional coefficients are introduced [8]. An example of coefficients modifying empirical formulas for semi-hard missiles applicable in one Member State can be found in Ref. [8].

Parts of an aircraft that are important for deriving the loading function due to the impact of an aircraft crash are illustrated in Fig. 2.

In order to evaluate the loading function due to the crash of a given aircraft, the following investigations need to be done:

— Clarification of site specific impact scenarios (e.g. impact in unprotected locations);
— Analyses of crash processes and derivation of representative loading functions.

Examples of predictions of loading functions for impacts of a military and a commercial aircraft are given in Appendix I.

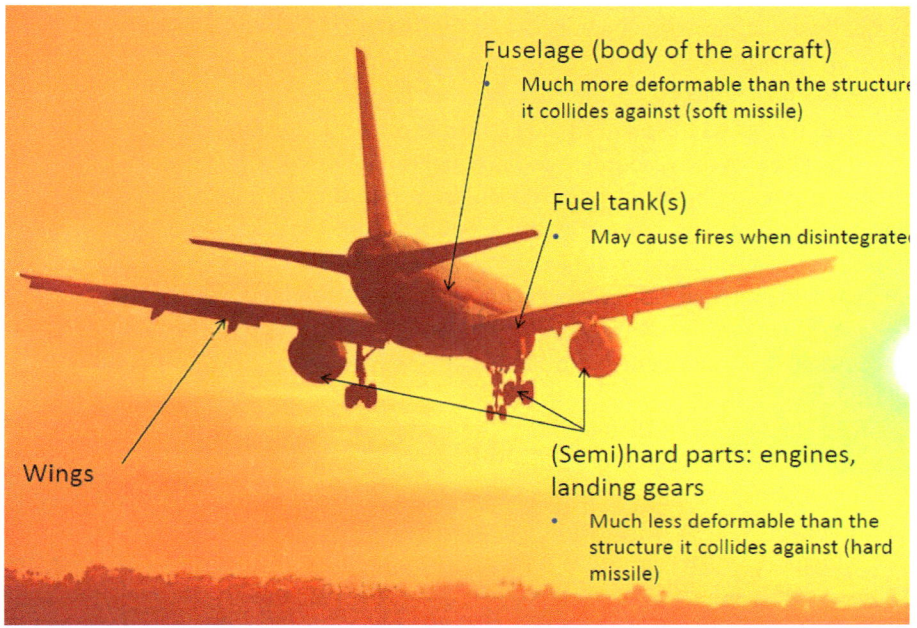

Fuselage (body of the aircraft)
• Much more deformable than the structure
  it collides against (soft missile)

Fuel tank(s)
• May cause fires when disintegrated

(Semi)hard parts: engines,
landing gears
• Much less deformable than the
  structure it collides against (hard
  missile)

Wings

FIG. 2. Aircraft parts classified on the basis of the loading they cause.

2.3. EXPLOSIONS

2.3.1. General considerations

An explosion is defined as a large scale, sudden release of energy. Explosions can be classified on the basis of their nature as physical or chemical events. In physical explosions, energy may be released, for example, from the catastrophic failure of compressed gas containers, whereas in the case of chemical explosions, the rapid oxidation of explosive materials is the main source of energy.

The effects of explosions on structures can be categorized as:

— Blast pressure loading;
— Blast generated missiles;
— Blast induced ground motion;
— Thermal loading.

This report focuses on blast pressure loading.

Blast pressure loading depends on the rate at which energy is transferred from the exploding material to the transmitting media. For chemical explosions, typically a distinction is made between 'deflagration' and 'detonation' [1, 2]. The difference is in the speed of the reaction process causing energy release through the exploding material. If the reaction moves through the explosive material at less than the speed of sound in the explosive material, the explosion is considered a deflagration. If the reaction speed is equal to, or greater than, the speed of sound in the explosive material, it is considered a detonation. In turn, a detonation can be 'confined' or 'unconfined'. A confined detonation takes place when the explosive material is surrounded by a casing having some significant strength, such as a metal shell. This report focuses on unconfined detonations.

With regard to the physical effects of detonations on a given structure, it is reasonable to distinguish between distant blasts, near field blasts and contact blasts.

In the case of a distant blast, the main loading effect is the overpressure caused by the incoming pressure wave, increased by wave reflection. Other loading effects, such as those due to drag caused by the 'blast wind', can also be important.

In the case of structures subjected to a near field or a contact blast, the shock wave originating from the detonation travels directly through the solid construction material causing a sharp increase in pressure, and loading conditions very different to the loads of a distant blast. For screening estimates, such structures may be assumed to have failed under the load condition of a near field blast. For detailed analysis, a methodology suitable for the analysis of effects of near field blasts needs to be used, including shock wave propagation through the solid material of the structure.

For the purposes of characterizing blasts, it is assumed that the energy of the explosion may be described adequately by an equivalent weight of a spherical or hemispherical charge of trinitrotoluene (TNT). Much data exists to correlate explosive materials to equivalent TNT charges (see, for example, table 2-1 of Ref. [9] and eq. (2-1) of Ref. [10].

It is, therefore, reasonable to provide a simple criterion to distinguish between distant, near field and contact blasts. Such a criterion can be provided based on the Hopkinson scaled distance Z [10]:

$$Z = \frac{R}{W^{1/3}} \tag{2}$$

where

R is the stand-off distance (the distance between the location of the explosion and the structure of concern) (m);

and W is the equivalent TNT mass for the explosive of concern (kg).

In blast equations, W is traditionally used for mass.

The contact, near field and distant blasts are defined based on the following limits on Z:

— Contact blast: $Z < 0.4$ m/kg$^{1/3}$.
— Near field blast: $0.4 < Z < 1.5$ m/kg$^{1/3}$.
— Distant blast: $Z > 1.5$ m/kg$^{1/3}$.

The near field blast region is the flame region, and blast loading includes very high pressures and temperatures. In the distant blast region, the loading is due to the blast wave. Other definitions also exist (see, for example, Ref. [11]).

The design methods based on average blast loads normally do not account for increased blast effects produced by a contact blast. Thus, a separation distance between the element and the explosive is to be maintained. This separation is measured between the surface of the element and the surface of either the actual explosive or the spherical equivalent, whichever results in a larger distance between the element's surface and the centre of the explosive. For the purpose of design, a separation scaled distance corresponding to $Z = 0.4$ m/kg$^{1/3}$ is recommended for a mass up to 227 kg in Ref. [9].

2.3.2. Distant blasts

2.3.2.1. Blast pressure waves

Given an estimate of the TNT equivalent charge and the configuration of the scenario, the following free air blast parameters can be derived. The typical pressure–time history of an air blast in free air is given in Fig. 3.

The incident wave is characterized by:

— Wave velocity U;
— Arrival time t_a;
— Peak positive pressure (also 'peak side-on overpressure') P_{so};
— Positive phase duration t_o;
— Peak negative pressure P_{so}^-;
— Negative phase duration t_o^-.

Positive and negative peak pressures are measured relative to ambient pressure p_o. The derived parameters are positive and negative phase impulses, which correspond to the areas integrated from each part of the pressure–time curve.

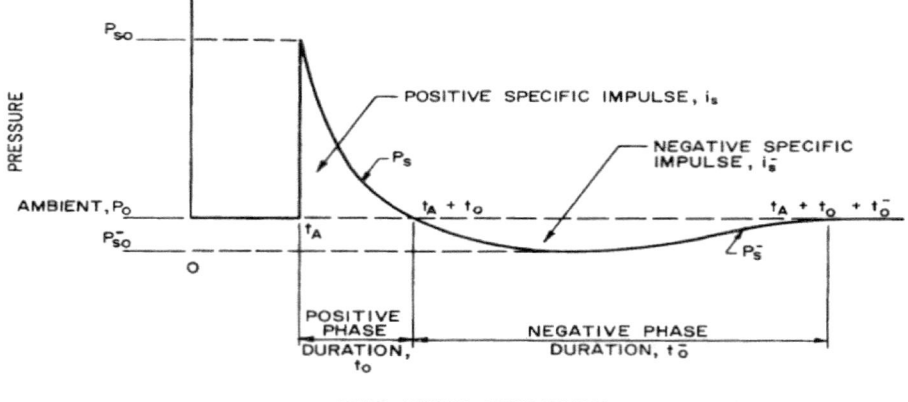

FIG. 3. Pressure–time variation of a spherical blast in free air [9].

The decay of the peak positive pressure is usually described by the modified Friedlander equation, which is a quasi-exponential approximation of the positive phase of the waveform of Fig. 3, accounting for the ambient pressure [12]:

$$P_s(t) = P_{so}\left(1 - \frac{t}{t_o}\right)e^{-\alpha\frac{t}{t_o}} \tag{3}$$

where α is a dimensionless parameter that describes the decay of the curve.

The peak overpressure–distance relation for an air blast in free air can be written as [12]:

$$P_{so} = P_0 \frac{808\left[1 + \left(\frac{Z}{4.5}\right)^2\right]}{\sqrt{1 + \left(\frac{Z}{0.048}\right)^2}\sqrt{1 + \left(\frac{Z}{0.32}\right)^2}\sqrt{1 + \left(\frac{Z}{1.35}\right)^2}} \tag{4}$$

where

Z is the Hopkinson scaled distance (m/kg$^{1/3}$) given by Eq. (2);

and p_0 is the ambient pressure.

The positive specific impulse or impulse per unit area i_s (in pascal seconds) of the blast wave is the time integral over the positive phase of the side-on overpressure:

$$i_s = \int_0^{t_0} P_s(t) \, dt$$

(5)

Inserting Eq. (3) into Eq. (5) and performing the integration gives:

$$i_s = P_{so} t_o \left[\frac{1}{\alpha} - \frac{1}{\alpha^2} \left(1 - e^{-\alpha} \right) \right]$$

(6)

In practice, the waveform in Fig. 3 can usually be replaced by a right triangle having the same values of peak pressure P_{so} and positive specific impulse i_s as the actual blast wave. The effective duration t_i of the triangular blast wave is calculated from:

$$t_i = \frac{2 i_s}{P_{so}}$$

(7)

Blast wave parameters are usually plotted as functions of Hopkinson scaled distance Z, as in Fig. 4. These parameters are based on empirically defined equations and should not be extrapolated beyond the ranges given in the literature.

2.3.2.2. Free air, free field and reflected blast waves

The blast parameters defined for a 'free air blast', remote from any reflecting surface, are commonly referred to as 'spherical air blast' parameters or 'spherical incident wave' parameters (Fig. 4). When the explosive is in contact with the ground ('hemispherical ground blast'), the parameter values given for a spherical air blast are modified to account for the reflection from the ground. This is usually accomplished by multiplying the TNT equivalent mass W by a factor of 1.8.

An explosion creates an incident free field wave moving forward with the wave front velocity U and the peak side-on pressure P_s on any surface perpendicular to the wave front (angle of incident shock $\beta = 90°$), as shown in Fig. 5. When the blast wave impinges onto a surface (angle of incident shock $\beta < 90°$), a new reflected blast wave is formed. The effect of this blast wave reflection is that the surface will experience a pressure larger than the incident

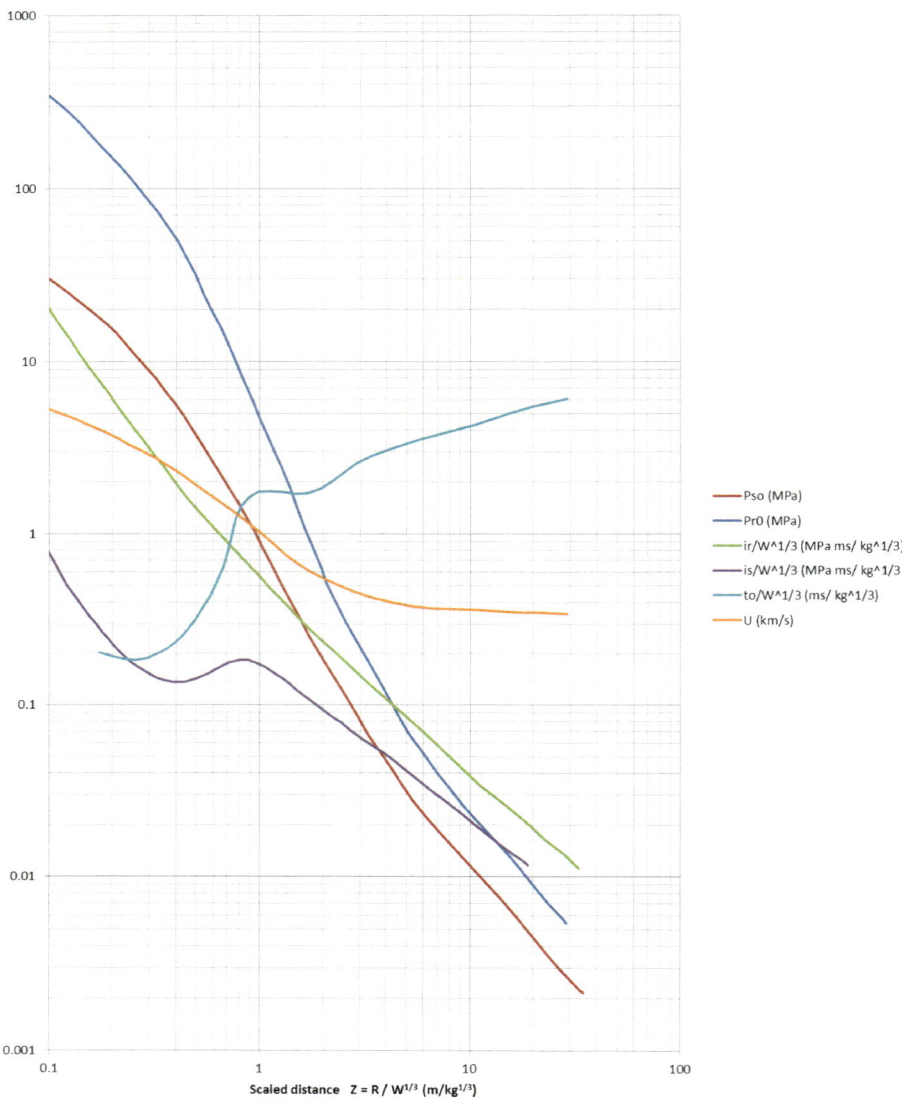

FIG. 4. *Blast wave parameters for spherical charges of trinitrotoluene (adapted from Ref. [9]).*

side-on pressure. This pressure is the 'face-on' pressure. The highest pressure peak occurs in the normal reflection ($\beta = 0°$).

When the blast wave is reflected from a large surface parallel to the wave front ($\beta = 0°$), the ratio of peak value of the reflected pressure P_{r0} to that of side-on pressure P_{so}, also called the reflection coefficient, can be computed as:

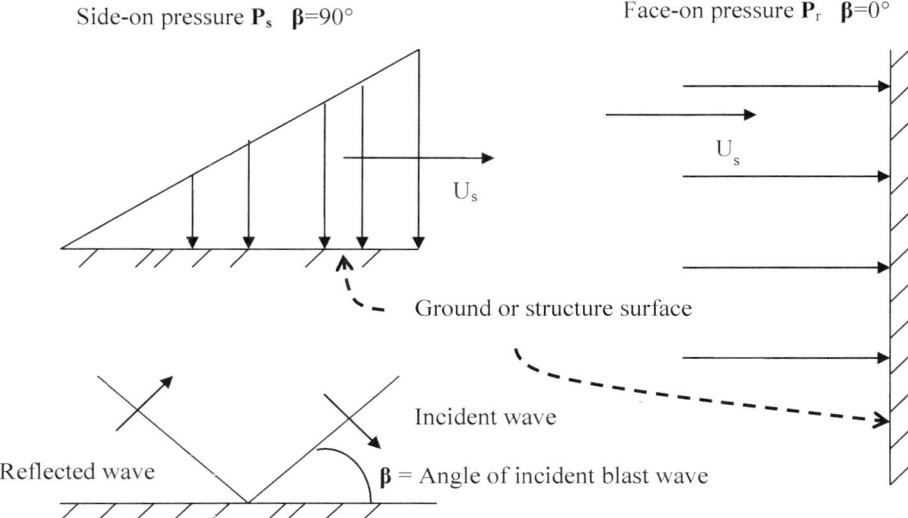

FIG. 5. Side-on and face-on pressures, and angle of incidence for a blast wave.

$$\frac{P_{r0}}{P_{so}} = 2 + \frac{6P_{so}}{P_{sc}} + 7p_o \qquad (8)$$

When the side-on pressure P_{so} exceeds 10 bars, the shock temperature is so high that air molecules dissociate and ionize. The assumption used in the derivation of Eq. (8), that air behaves as an ideal gas, is no longer valid. A correlation for the values of P_{r0}/P_{so} corresponding to normal reflection, calculated assuming air dissociation and ionization, can be found in Ref. [10]. The correlation in Ref. [10] is usually replaced by a simpler one [13]:

$$\frac{P_{r0}}{P_{so}} = 4\lg P_{so} + 1.5 \qquad \frac{P_{r0}}{P_{so}} < 14 \qquad (9)$$

where P_{so} is introduced in bars.

When the angle β between the blast wave plane and the surface increases, the reflected peak overpressure decreases (oblique reflection). When the angle reaches a relatively high value, the shock front no longer bounces away directly from a surface, but is instead deflected so that it spurts along over the surface. This is called a Mach stem regime. The overpressure in the Mach stem regime is significantly lower than that of the normal and oblique reflections. In most cases, the minimum angle which permits a shock front to form a Mach stem is about 39°. The dependences of reflection coefficient on the angle β and on the peak value of the side-on pressure P_{so} are depicted in Fig. 6.

The method for calculating the reflection coefficient for oblique and Mach reflection is presented, for example, in Ref. [12]. To take into account the wave reflection in the structural design, the peak value of the side-on pressure P_{so} given by Eq. (3) is multiplied by the reflection coefficient, which is a function of the angle of the incident wave (Fig. 6).

Curves for the scaled specific impulse of the reflected blast $i_r/W^{1/3}$ versus the angle of the incident blast wave are presented in Fig. 7 for hemispherical blasts. The actual time dependence of the reflected blast is sometimes replaced by a right triangle with the peak pressure P_{r0} and specific impulse i_r. The effective duration (base of the triangle) is calculated as in Eq. (7).

2.3.2.3. Blast loading of structures

The interaction of the free field blast waves with SSCs is a complex phenomenon. This is true even without considering the interaction between adjacent structures and components, typically present on a nuclear power plant site, that serve to shield or to amplify the blast loading conditions.

When the blast wave strikes a parallelepiped building, transient blast loads act on the walls and the roof (Fig. 8). The pressure time histories to be applied to the front wall, side walls, roof and rear wall are different to each other.

When a closed rectangular structure is loaded by a blast wave from a detonation at a distance sufficiently great compared to the structure dimensions, the blast can be approximated by a plane wave, usually assumed to be parallel to the front wall. Figure 8 shows, schematically, the behaviour of the blast when it meets a building with solid walls. Four stages are differentiated in Fig. 8:

(a) The wave front is still in front of the building and is not yet disturbed.
(b) The wave front has reached the building. Reflection takes place and a rarefaction wave is formed.
(c) The blast wave envelops the structure.
(d) The wave front has passed the building.

Initially, the front wall is loaded by the reflected pressure. Owing to the disturbance of the incident blast wave, major pressure differences develop at the edges of a reflecting surface. As a consequence, a rarefaction wave begins to progress, from the edges, on the front wall (Fig. 8). As the blast wave front moves forward (Fig. 8), the reflected overpressure on the front wall drops rapidly due to this rarefaction wave to the so-called stagnation pressure P_{stag} consisting

(a)

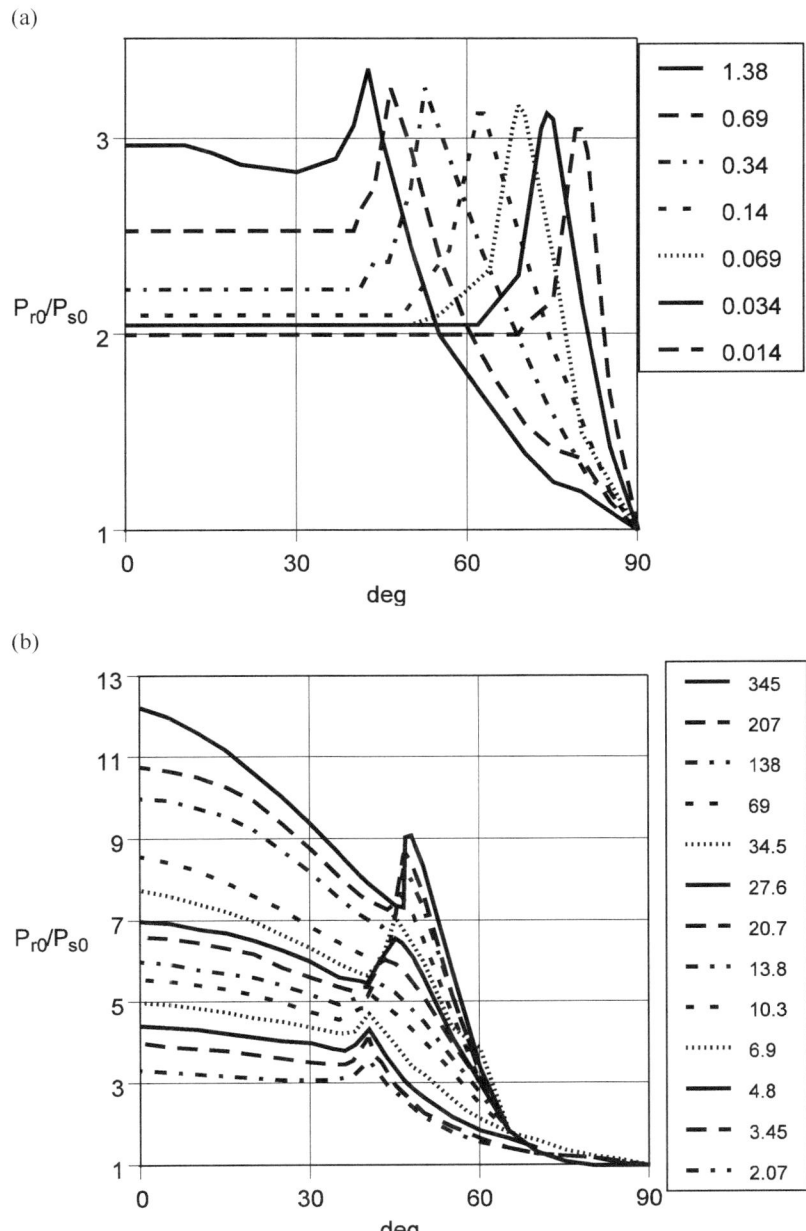

FIG. 6. *Reflection coefficient P_{r0}/P_{so} versus angle of the incident blast wave for several values of peak side-on overpressure P_{so} in SI units (adapted from Ref. [9]). (a) Peak side-on overpressure P_{so} less than 2 bars. (b) Peak side-on overpressure P_{so} greater than 2 bars.*

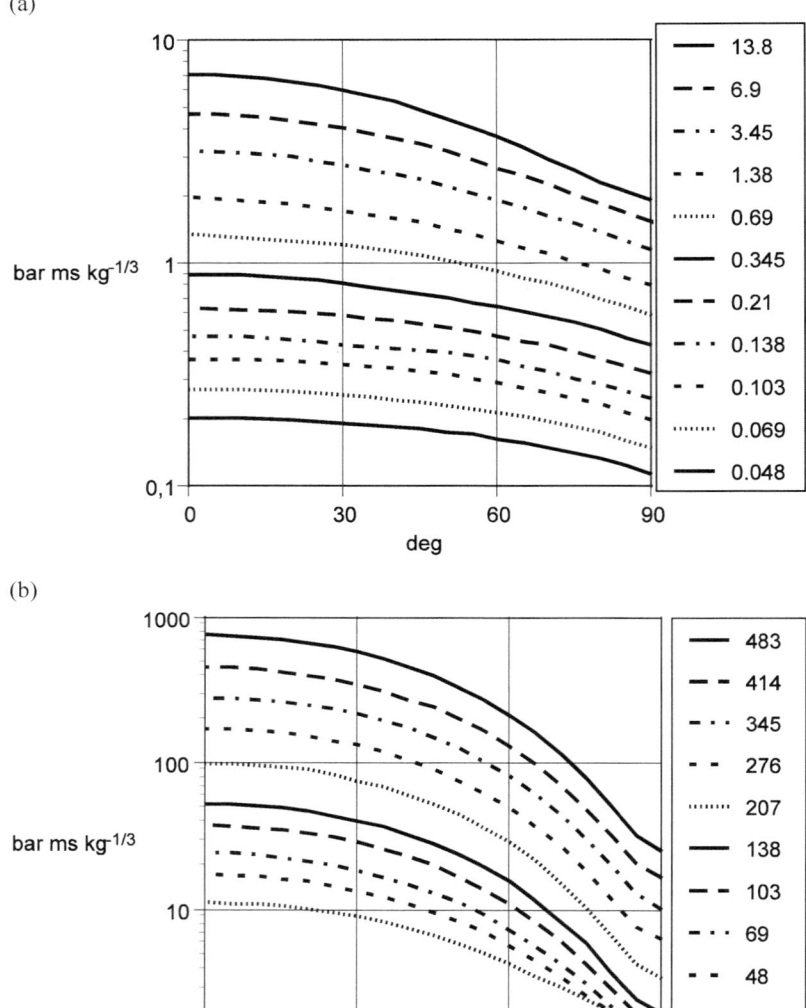

FIG. 7. *Scaled reflected specific impulse of the reflected blast* $i_r/W^{1/3}$ *versus angle of the incident blast wave for several values of peak side-on overpressure* P_{so} *in SI units (adapted from Ref. [9]). (a) Peak side-on overpressure* P_{so} *less than 20 bars. (b) Peak side-on overpressure* P_{so} *greater than 20 bars.*

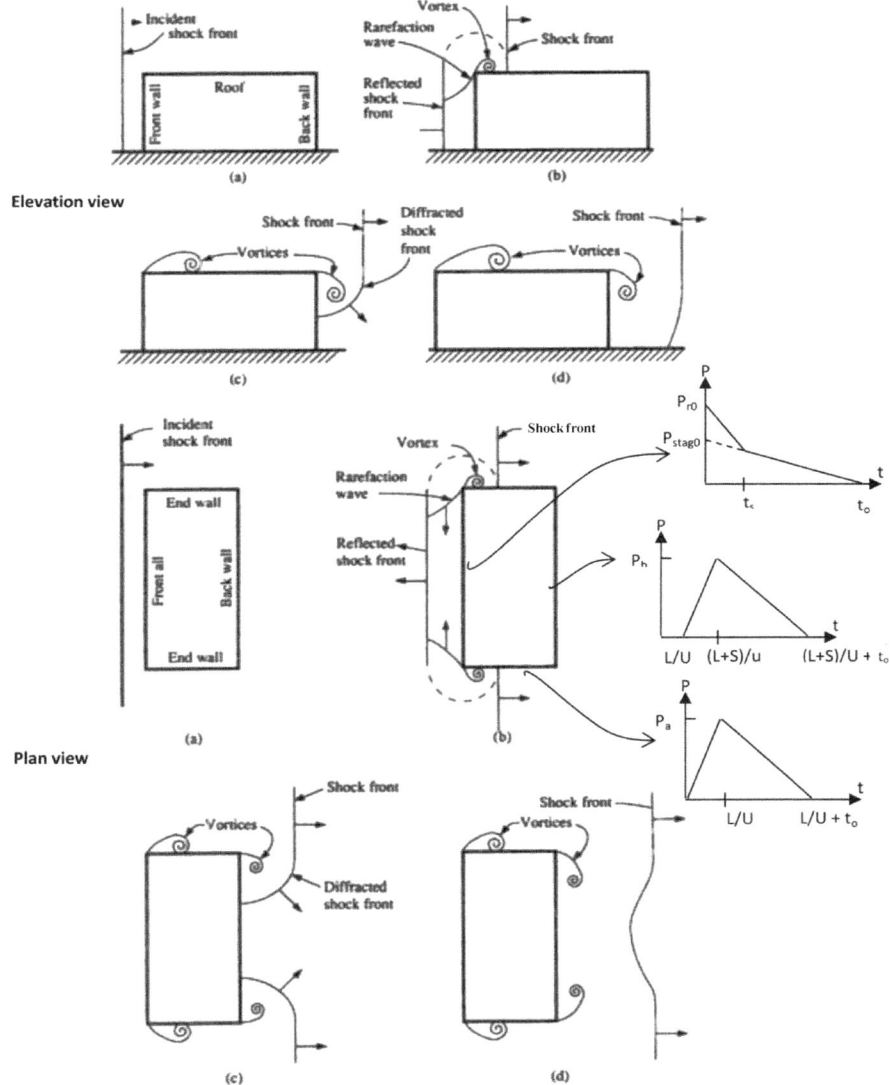

FIG. 8. Schematic representation of a blast wave passing a rectangular building (adapted from Ref. [14]).

of the side-on overpressure P_s plus the drag force $C_D Q$, which corresponds to the movement of air caused by the blast (blast wind). Here, C_D is the drag coefficient, which can be taken to be equal to 1.0 for the wall facing the explosion and to –0.4 for other walls and the roof of the building.

The height of the building is denoted by H and the width by B. The characteristic dimension of the front wall S is defined as the smaller of H and $B/2$. Laboratory studies have shown that the stagnation pressure is reached on the front wall at time $t_s = 3S/U$, where U is the wave front velocity. Time t_s should not be taken to be greater than t_o, the positive phase duration.

Hence, the average pressure on the front wall decays from P_{r0}, when the shock front reaches the wall, to P_{stag} at t_s:

$$P_{stag} = P_s + C_D Q = P_s + C_D \frac{2.5 P_s^2}{P_s + 7 p_o} \tag{10}$$

and then it goes further to zero at t_o, the positive phase duration.

The drag force in Eq. (11) can be the dominant blast effect for open frame structures, but it is usually negligible for buildings closed by solid walls, such as the ones normally found in nuclear power plants. If the pressure–time history is idealized as a bilinear curve (Fig. 8), the specific impulse on the wall can be approximated by:

$$i_w = \frac{1}{2}\left(P_{r0} - P_{stag0}\right) t_s + \frac{1}{2} P_{stag0} t_o \tag{11}$$

The blast loads on the rest of the walls and the roof due to a passing blast wave front depend on the location and time of the blast. Usually, however, it is enough to calculate the average pressures on them. The length of the building is denoted by L. The loading of side walls and the roof begins when the wave front passes the front wall. The loading attains its peak value when the wave front has passed the building, that is, at $t_r = L/U$, where L is the building length. The peak value of the average pressure can be obtained as:

$$P_a = C_e P_{so} + C_D Q_o = C_e P_{so} + C_D \frac{2.5 P_{so}^2}{P_{so} + 7 p_o} \tag{12}$$

where C_e is a reduction factor that depends on the relative size of the building with respect to the wavelength $L_w = U t_o$. The factor C_e varies from 0.2, for $L_w/L = 0.35$; 0.4, for $L_w/L = 0.85$; 0.45, for $L_w/L = 1.0$; 0.7, for $L_w/L = 2.0$; 0.8, for $L_w/L = 3.0$; and 0.9, for $L_w/L > 5.0$. Additional values for C_e can be obtained from Ref. [9]. After the peak, the average pressure decays to zero at time t_o after the peak (Fig. 8).

The average pressure on the back wall begins at L/U and attains its peak value P_b at $t_r = (L + S)/U$:

$$P_{\text{b}} = P_{\text{so}} + C_{\text{D}}Q_{\text{o}} = P_{\text{so}} + C_{\text{D}}\frac{2.5P_{\text{so}}^2}{P_{\text{so}} + 7p_{\text{o}}} \qquad (13)$$

Thereafter, it decreases to zero, which is attained at $t_{\text{r}} + t_{\text{o}}$ [15] (Fig. 8). The waveforms of the average loads on building walls and roof are presented, for example, in Refs [15, 16].

The pressure difference between the front and back walls will have its maximum value when the blast front has not yet completely engulfed the building (Fig. 8). Such a pressure differential will produce a force tending to cause the structure to deflect and, thus, to move bodily, usually in the same direction as the shock wave. This force is called diffraction loading because it operates while the shock wave is being diffracted around the building [14, 16].

The interaction of an ideal plane blast wave with an upright cylinder which has an axial dimension much larger than the diameter D (e.g. a stack), when the wave propagates in a direction perpendicular to the cylinder axis, can be described in terms of resultant forces on the cylinder. When the blast wave reaches the cylinder, the load on the cylinder appears as a force that increases with time from zero, when the shock front arrives, to a maximum at $D/2U$, when the front has propagated one radius. The maximum average pressure acting on the projected area of the cylinder (i.e. D per unit length) is approximately twice the maximum side-on overpressure, $2P_{\text{so}}$. The load then decays in an approximately linear manner to the value of $P_{\text{s}} + C_{\text{D}}Q$ at time $2D/U$. The drag coefficient for the front surface of the cylinder can be taken as 0.8. Subsequently, the load decays to zero.

Loading on the sides commences immediately after the shock front strikes the front surface, but the sides are not fully loaded until the front has travelled distance D, i.e. at time D/U. The average pressure on the projected area of the sides is approximately P_{s} at time D/U. Complex vortex formation then causes the average pressure to drop to a minimum at time $3D/2U$. The value of the minimum is about one half the value of P_{s} at this time. The average pressure on the side then rises until time $9D/2U$ and subsequently decays. The drag coefficient for the side face is –0.9 [16].

The shock wave begins to affect the back surface of the cylinder at time $D/2U$ and the average pressure on the projected surface gradually builds up to half the value of P_{s} at $4D/U$. The average pressure continues to rise until it reaches a maximum at a time of about $20D/U$, which is $P_{\text{s}} + C_{\text{D}}Q$ at time $20D/U$. The average pressure on the back then decays. The drag coefficient for the back face is –0.2 [16].

The preceding discussion concerns the average values of the loads on the various surfaces of a cylinder, whereas the actual pressures vary continuously from point to point. Consequently, the net horizontal loading cannot be determined accurately by the simple process of subtracting the back loading from the front loading. It is necessary to sum the horizontal components of the loads over the two areas and then subtract them [16].

The waveforms of the average loads on the various surfaces of a cylinder are presented in Ref. [16].

In order to analyse cylindrical structures with values of D which are not much smaller than the length, an analysis with average pressures may not be accurate enough. For such structures, Ref. [16] presents a method by which the time dependence of the blast load at different points of the cylinder surface can be approximated.

In Ref. [17], the non-linear response of a reinforced concrete nuclear containment structure under blast loading is simulated. The reinforced concrete shell is composed of cylindrical and spherical parts of constant thickness. The cylindrical part has an inner diameter of 39.6 m, an outer diameter of 42 m and a height of 46 m. The nuclear containment shell was subjected to surface blast loading of 10.0, 12.5, 15.0, 17.5 and 20.0 t of TNT at a stand-off distance of 100 m.

When a charge is detonated at a distance comparable to building dimensions, the blast wave front cannot be approximated by a plane wave. In this case, the nearest wall or walls have to be divided into surface elements and the loading of each element has to be calculated. The distance from the charge to each element is calculated and converted to a Hopkinson scaled distance. Then, the parameters of the side-on pressure wave (t_a, P_{so}, t_o, α) at this scaled distance are calculated from the correlations given in Ref. [12]. The positive specific impulse i_s is calculated from Eq. (6) and effective duration t_i from Eq. (7). The peak value of the reflected pressure P_r is calculated from the values of P_s and the angle β. For normal reflection, Eq. (8) or Eq. (9) is used, and for oblique reflection, the method presented in Ref. [12] or a procedure based on the parameterization of Figs 6 and 7 is applied. In practice, these calculations are done with a computer code which also produces an input file for the structural analysis program.

When buildings cannot be considered as isolated, the blast loads are affected by the presence of adjacent structures. They can be either reduced due to shadowing by other buildings or augmented due to reflection and channelling of the blast wave. A limited number of experiments involving small scale models have been conducted to validate numerical simulations of blast wave–multiple structure interaction. The usual approach is to simulate blast propagation with numerical methods incorporating computational fluid dynamics (CFD) techniques, so-called hydrocodes.

Reference [18] presents simulation results for two buildings, the first of which partially shadows the second one, and for a blast wave propagating along a street. Reference [19] reports a study in which experiments were done with a small scale model consisting of four concrete cubic boxes (dimensions: 2.3 m) positioned symmetrically at a distance of 2.3 m apart. Explosive charges were placed at four different positions and a total of 25 pressure gauges were used to register the pressure–time relations at various locations. Twenty gauges had a fixed position on the concrete boxes. The remaining five gauges were used to register the free field pressure. The Hopkinson scaled distance ranged from 1.5 to 13 m/kg$^{1/3}$ and from 1.0 to 8.5 m/kg$^{1/3}$ for the 0.4 and 1.6 kg TNT charges, respectively.

The tested configurations were simulated numerically and the simulations provided very good agreement with the measurements [17]. The results were used to derive a simple engineering method in which several blast waves are superposed and every wave is adjusted with regard to diffraction. The influence of reflected pressure is not considered, that is, the wave characteristics are based on incident pressures only, and there is no attempt to incorporate any confinement effects.

Using this method, the resulting pressure–time relation for a particular point is determined according to the following steps:

(a) All blast wave paths of interest for the point are derived and the total distances from the centre of charge to the point are calculated.
(b) The corresponding Hopkinson scaled distances are used to calculate the respective parameters of the positive and negative phase of the incident blast waves.
(c) The pressure–time histories for all wave paths are calculated.
(d) The pressure–time histories for each wave are multiplied by the diffraction coefficient and then summed. The approximate average value of the diffraction coefficient was 0.6^n, where n is the number of corners the wave diffracts around.

2.3.3. Vapour cloud explosion

A vapour cloud explosion is the outcome of a process in which flammable material is first released into the atmosphere and dispersed in air, forming a vapour cloud. Then, after some delay, the flammable portion of the vapour cloud is ignited, producing an explosion [20]. Consequently, these three conditions have to be met for a vapour cloud explosion to take place [20]:

(a) A release of flammable material into a confined/congested area;
(b) A delayed ignition, so that the formation of a flammable mixture with air has occurred;
(c) An ignition source with enough energy to ignite the fuel–air mixture.

A 'confined' area means the presence of solid surfaces able to prevent displacement of the unburned gases and a flame front in one or more dimensions. A 'congested' area means the presence of obstacles in the path of the flame which generate a turbulent flow. Turbulence tends to fold and to wrinkle the flame, which, in turn, increases the flame surface area and combustion rate. Hence, a number of closely spaced obstacles along the path create favourable conditions for a vapour cloud explosion.

The overpressure in a vapour cloud deflagration depends on the combustion rate: a low flame speed produces a small overpressure and larger flame speeds produce higher overpressures. Detonations are very unlikely in accidental vapour cloud explosions.

In the 1970s, before the mechanisms of overpressure generation in vapour cloud explosions were fully understood, blast predictions were made with the TNT-equivalency method. An equivalent charge mass of TNT was calculated by multiplying the available heat of combustion in a vapour cloud by a conversion factor. The theoretical upper limit for the conversion factor is approximately 40%. Reported values of the conversion factor, deduced from damage observed in many explosion incidents, range from 1% to 10%. Apparently, only a small part of the total available combustion energy is generally released in actual explosions.

In the 1980s, after investigating damage patterns in major accidental vapour cloud explosions, it was found that there is very little correlation between the total available combustion energy and the effects of the explosion. Blast parameters are dependent on the size and nature of partially confined and congested areas within the cloud [20]. The remaining portion of the cloud (i.e. not in these areas) containing a flammable vapour–air mixture burns out slowly without contributing significantly to the blast. These are the underlying considerations in the multi-energy method.

The basic tool of the multi-energy method consists of blast curves that present blast parameters as a function of dimensionless Sachs scaled distance $\bar{R}$ [10]:

$$\bar{R} = R\left(\frac{p_o}{E_b}\right)^{\frac{1}{3}} \tag{14}$$

where

R is the stand-off distance (m);
p_o is the ambient (atmospheric) pressure (Pa);

and E_b is the combustion energy of the cloud (J) ($\approx$volume of the partially confined and congested region (m^3) multiplied by 3.5 MJ/m^3 for hydrocarbon clouds).

Figure 9 represents the blast characteristics of a hemispherical fuel–air cloud of radius R_o on the ground surface. Dimensionless side-on peak overpressure and positive phase pressure wave duration are presented as a function of the Sachs scaled distance from the blast centre. The initial blast strength in Fig. 9 is represented by a number ranging from one (very low strength) to ten (detonative strength). In addition, this figure gives a rough indication of the blast wave shape. The procedure for applying the multi-energy method to model a vapour cloud explosion blast is presented in Ref. [20].

For extended and quiescent parts of the vapour cloud, a minimum strength of one is assumed. For parts in low intensity turbulent motion, a strength of three is assumed [20].

The most difficult step in using Fig. 9 is estimating the strength of the sources of the explosion. At a distance greater than $10R_o$, the pressure wave is more or less independent of initial strength for values of six (strong deflagration) and above. A very conservative estimate is to assume the maximum strength of ten. However, a strength of seven more accurately represents actual experience. Furthermore, for Sachs scaled distances greater than 0.9 (side-on overpressures below about 0.5 bars), there is no difference in predicted overpressure for source strengths ranging from seven to ten.

If the approach above yields unacceptably high overpressures, the initial blast strength may be determined from experimental data on gas explosions. One way of doing this is to apply the congestion assessment method, which consists of simple correlations derived from a large number of experiments. Another possibility is to apply numerical simulation by use of advanced CFD codes [20, 21].

2.3.4. Contact and near field blast

In the case of a near field or contact explosion, the shock wave originating from the detonation travels directly through the solid construction material causing a sharp increase in pressure, and loading conditions are very different to the loads of a distant explosion. Close to the detonation charge, the shock temperature is so high that air molecules dissociate and ionize. Air no longer behaves as an ideal gas. As a result of this phenomenon, the pressure peak values become considerably higher. The pressure effects/blast loading, which are publically available, are given in Table 2.

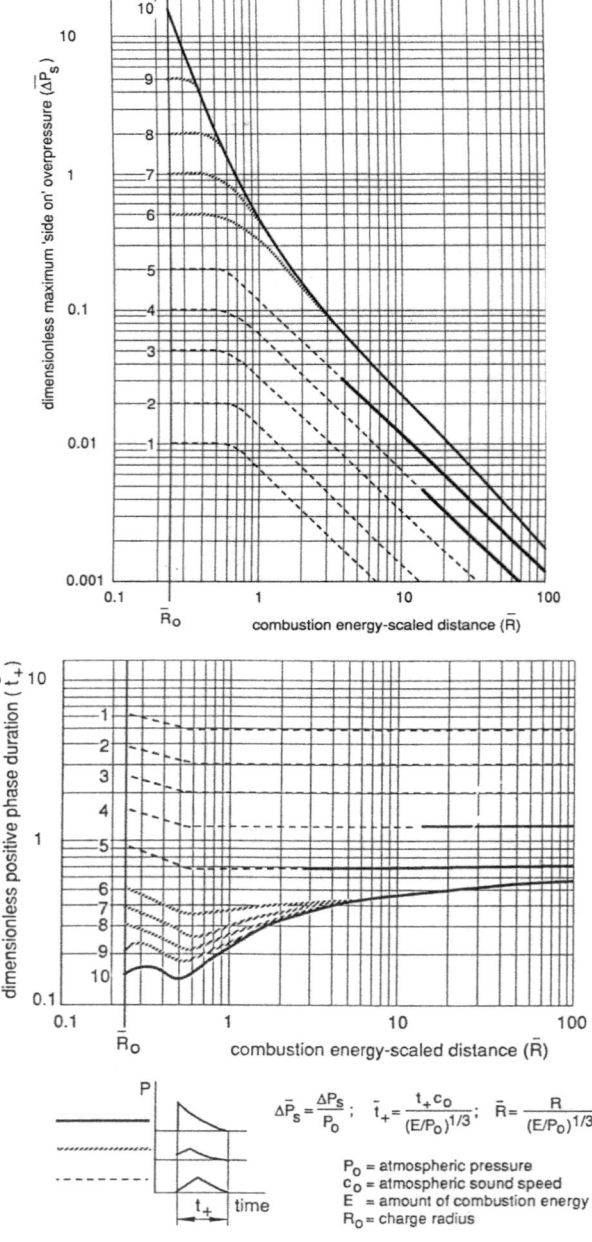

FIG. 9. *Multi-energy method: side-on, blast overpressure and duration curves [20].*

TABLE 2. EXAMPLES OF PRESSURE EFFECTS

Mass (kg)	Stand-off (m)	Side-on pressure (kPa)	Incident impulse (Pa · s)	Duration (ms)
10	2	202	252	2.5
75	10	193	483	5.0
1000	20	284	1345	8.5

2.4. FIRE

2.4.1. Procedure for fire consequence analysis

Many human induced external events may result in an external or internal fire. Examples of such events are aircraft related fireballs and explosions of in situ storage of flammable liquids and gases. As most safety related SSCs are usually located inside plant buildings, it may be assumed that internal fires dominate the risk of damage to SSCs. External fires can, however, influence the conditions inside the buildings or even spread inside, igniting internal fuels.

The process of analysing the consequences of a fire is illustrated in Fig. 10. The individual tasks of the process are described below:

(a) 'Scenario setting' consists of choosing between the physical impact scenario and the fire scenario. The fire scenario specifies the materials that contribute to the fire, the environmental conditions of fire development and the role of automatic and manual firefighting measures. Scenario selection takes into account the deterministic or probabilistic nature of the analysis. Additional guidelines are provided in Section 2.4.2.

(b) The 'physical damage footprint' is determined by analysing the mechanical response of the structures under the load associated with the scenario. The extent of the physical damage defines the part of the plant that can be affected by the fire directly after the event.

(c) When the physical damage is determined, the fire analysis will focus either on the internal fires, caused by the entrance of flammable fuels or a fireball inside the building, or on the external fires.

(d) If the initial fireball or fire cannot enter the building, an analysis of the external fires will be carried out, as explained in Section 2.4.3.

(e) The performance of the external construction will be assessed (Section 5.5.1) to determine whether the influence of the fire will reach the

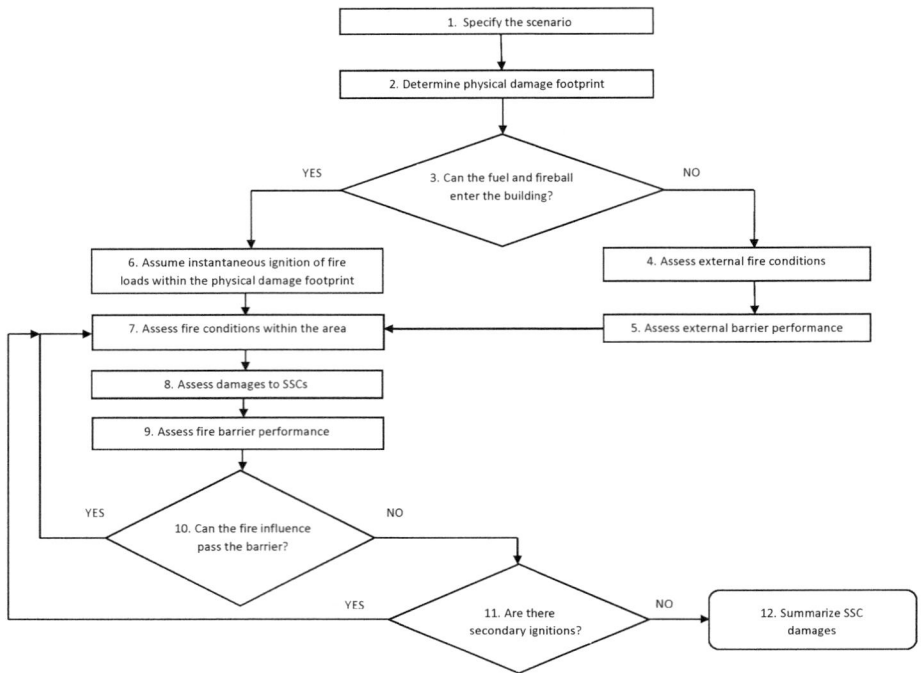

FIG. 10. Flow chart of the analysis of the consequences of the fire. SSC: structures, systems and components.

interior of the plant (e.g. by convective heat through the external wall or by damaging the external wall so that flames extend into the interior of the building).

(f)   If the external fire does enter the building through the physical damage zones, it can be assumed that all or some of the combustible materials within the fire area will ignite immediately, along with any fuel that may have been outside the building or part of the aircraft or other missile.

(g)   Internal fires are analysed (Section 2.4.4).

(h)   The internal fire analysis is followed by the assessment of SSC damage (Section 5.5.2).

(i)   The performance of the fire barriers is assessed (Section 5.5.1).

(j)   If the internal fire barriers fail, the fire will spread to the next fire area, and the steps of internal fire analysis, SSC and fire barrier performance will be repeated.

(k)   If there are additional ignition sources, the steps of internal fire analysis, SSC and fire barrier performance will be repeated.

(l)   When further spreading of the fire influence is not foreseen, the information on the SSC damage is summarized to evaluate the capacity to perform the safety functions.

It should be noted that the timescales of the different phases of this analysis are very different:

(a)   The physical damage due to the impact usually takes place in less than a second.
(b)   An external fireball and possible internal fireball have a timescale of a few seconds.
(c)   A fire involving large internal fire loads may have a duration of hours.
(d)   Failure of SSCs under direct flame contact or severe thermal radiation takes from a few seconds to minutes.
(e)   Fire-induced failure of fire barriers generally takes tens of minutes or hours.

2.4.2.   Fire scenario

2.4.2.1. Main assumptions

A fire scenario represents a unique combination of events and circumstances that influence the outcome of a fire, including the impact of fire safety measures (see chapter 11-5 of Ref. [22]). Development of fire scenarios needs to take the following characteristics into account:

— Building characteristics, including the layout, materials and interconnections among compartments;
— Fuel loads and types of combustible material;
— Function in buildings;
— Passive fire protection systems;
— Detection and suppression systems;
— Fire department actions.

Conventional fire events are usually covered by the fire protection design of the facility. However, fire events which are not likely to be effectively covered solely by the facility protection systems and which may lead to extreme loading conditions need to be addressed individually. The boundary conditions with respect to fire events following an external event are usually significantly different compared to conventional fires that are analysed as part of the fire hazard analysis covered in Ref. [3]. For instance, fires resulting from an aircraft crash or explosion may be further challenging due to variables such as:

— A significant increase of ignition sources, including ignition sources directly related to the impact (fires, sparks, etc.) and those caused by secondary effects such as electrical surges within the building;
— An increase of fire loads outside and/or inside the buildings (generally fuel from the plane);
— The loss of fire protection systems outside the facility buildings, for instance, loss of water supply, pump stations, fire brigade installations or vehicles;
— The loss of automatic firefighting systems inside the facility buildings due to the impairment of water, foam or gaseous fire suppression systems;
— The loss of one or more facility buildings and their functions;
— The loss of off-site power in combination with the loss of SSCs;
— Weakening of fire barriers;
— Delayed manual response due to impaired or blocked access to the impact site or more than one fire event external and/or internal to the building;
— The loss of firefighting capabilities due to radiological effects.

The relevance of the fire scenarios needs to be evaluated from the viewpoint of their likelihood and severity. The most severe fire scenarios assume that entire buildings are on fire, leading to the following possible situations:

— A safety related building is totally involved in a fire and a second event occurs (e.g. loss of off-site power or loss of emergency power station); or
— Several safety related buildings are simultaneously involved in a fire due to fire spread and several fire sources.

If a fire occurs simultaneously inside and outside a building, it can be assumed that the predominant impact to the safety systems will result from the interior fire. However, the analysis of external fires is not to be ignored because external fires can challenge the separation and redundancy of the safety systems by extending the fire effects beyond the compartments affected by the interior fires. During an aircraft crash, the plane fire loads that are dispersed inside or outside the buildings may burn as fireballs or pool fires, or both. Explosions in certain areas of the affected buildings cannot be excluded without detailed investigations. Fireballs outside a building may cause fires and fire spread in adjacent buildings due to radiation effects. Pool fires outside a building may lead to local, thermally induced failure of structures or internal components or systems. Flowing liquid fuel can spread the fire to underground cable channels, to supply shafts of heating, ventilation and air conditioning systems or elsewhere. Fires outside a building can spread smoke to the air intakes and, therefore, challenge the operations inside the plant even if the fire itself does not spread inside.

The main element of the fire scenario is the 'design fire'. The design fire describes ignition time and place, how fast it develops, the peak heat release rate and the duration of the fire. Usually, the design fire is prescribed in terms of its heat release rate. Additionally, the design fire can, for instance, prescribe the production of toxic species and smoke.

Figure 11 shows schematically how an analysis of a fire scenario proceeds. Design fires inside of buildings might be ventilation controlled, so that the ventilation affects the development of the fire. The fire might not be able to develop freely due to active fire safety measures. Some kind of iterative

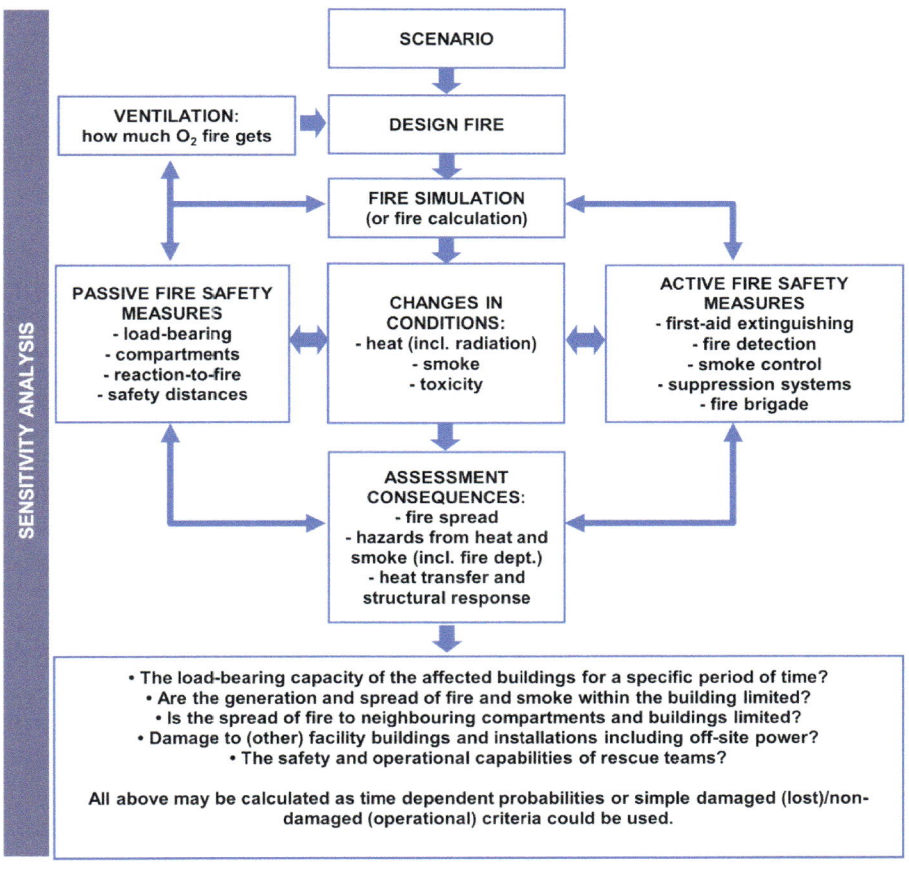

FIG. 11. A schematic diagram showing the options for a simulation of a fire scenario [23].

procedure might be needed, as can be seen in Fig. 11 with the different arrows that show the dependencies present in a fire scenario. The specific questions that need to be answered depend on each fire scenario; those given in the figure are for illustration purposes only.

2.4.2.2. Fuel loads

Fuel loads are the major component defining fire scenarios. Fuel loads consist of materials that can contribute to the combustion reactions during a fire. The most important characteristics of fuel materials are:

— The total mass available for combustion;
— Phase (condensed, liquid or gaseous);
— Heat of combustion;
— Availability for combustion (mass loss rate, flow rate, surface area);
— Ignition and flame spread characteristics.

The combustible fuels within the nuclear power plant are usually known from the fire hazard analysis, which is a part of the fire protection design concept [3]. In the context of human induced external events, the initial fires challenging the safety of the plant are usually extreme in power, geometrical size and growth rate. Some examples of relevant fuel loads are the fuel carried by an aircraft, or hydrocarbons carried by a tanker car or train. Small human induced fires outside the plant seldom lead to a loss of safety functions or ignition of internal fuel sources, and their effects are covered by the fire safety design principles.

The most significant means of transporting a large mass of fuel in the vicinity of a nuclear power plant are vehicles carrying combustible liquids or gases and commercial aircraft. The expected fuel mass carried by the vehicle needs to be based on the type of traffic in the vicinity of the nuclear power plant. For aircraft, the fuel mass can be estimated from the aircraft type, which can be divided into three groups. The estimated fuel loads of different types of passenger plane have been compiled in Table 3. The table does not cover the heaviest of the currently used planes, such as the Airbus A380, for which the maximum take-off weight is 560 t and the fuel mass 249 000 kg.

Table 3 indicates that the contribution of jet fuel to the total fuel loads is between 70% and 90%. It is, therefore, sufficiently precise to assume, in calculations, that the amount of jet fuel at take-off is the relevant fuel load for fire investigations.

The internal fuel loads are considered in the fire safety concept of the nuclear facility by implementing the defence in depth approach for fire safety. At the first

TABLE 3. FUEL LOADS OF PASSENGER PLANES AT TAKE-OFF MASS (kg) AND ENERGY (GJ) [24]

Aircraft type	Fuel (kg / GJ)	Lining seats (kg / GJ)	Luggage clothes (kg / GJ)	Combustible materials (kg / GJ)
Group A[a] <100 t	20 000 / 864	4 500 / 180	4 800 / 144	29 300 / 1 188
Group B[b] <200 t	56 000 / 2 440	8 800 / 350	8 400 / 250	73 200 / 3 040
Group C[c] <400 t	182 000 / 7 860	12 000 / 480	12 000 / 360	206 000 / 8 700

[a] Group A: light weight passenger planes such as the Boeing 737 and Airbus A320. The average number of passengers in a plane is about 160 and the take-off weight is below 100 t.
[b] Group B: medium weight planes such as the Boeing 757 and Airbus A300. The maximum number of passengers is 280 and the take-off weight is below 200 t. The Boeing 757 is the lightest plane in this group.
[c] Group C: heavy weight planes with take-off weights between 230 and 400 t. Heavy weight planes, such as the Airbus A340 and Boeing 747, provide about 400 passenger seats.

level of defence in depth, the ignition of fires is avoided by keeping the possible ignition sources away from the fuels and by choosing materials that are difficult to ignite. In the plant fire hazard analysis, a failure of the first defence in depth level is assumed. Next, the spread of fires and their consequences in areas with SSCs is reduced by the installation of fire detection and fire suppression systems. Furthermore, each building has a structural fire protection concept, which means that the building is subdivided into fire areas or fire compartments. The structural elements of the building and each compartment are designed according to a fire resistance classification system. In addition, each fire compartment can be divided further into fire cells according to the internal distribution of fire sources and threatened SSCs.

Fuel loads are quantified in terms of the total amount of thermal energy available for combustion and the heat release rate. The amount of thermal energy released in the fire is a product of the fuel mass M and the fuel heat of combustion H_c:

$$Q = H_c M \tag{15}$$

The values of H_c for many fuels are listed in chapters 1-5 and 3-4 of Ref. [22]. For most hydrocarbons in a liquid or gaseous state, H_c is between 40 and 50 MJ/kg. For solid fuels, values between 15 and 40 MJ/kg can be found.

The potential hazard of internal fires can be measured using the concept of fire load, which is the amount of combustion energy (in megajoules) that can be released from the materials inside a specific area or fire compartment. The fire loads are usually considered in relation to the specific floor area. In those cases, the fire loads are called fire load densities (or fire loads per unit area):

$$q = \frac{Q}{A} \tag{16}$$

where

q is the fire load density (or fire load per unit area) (MJ/m^2);

and A is the fire compartment area or reference floor area (m^2).

The most important boundary condition of the fire analysis is the fire's time dependent heat release rate (in megawatts). The heat release rate produced by an individual fire source is calculated as:

$$\dot{Q}(t) = \chi H_c \dot{M}(t) \tag{17}$$

where

χ is the combustion efficiency;

and $\dot{M}(t)$ is the fuel mass loss rate.

Factor χ accounts for the incomplete combustion which is typical for all fires. It depends strongly on the burning conditions and is, therefore, difficult to prescribe. In most analysis, it can be assumed that $\chi = 1$. The mass loss rate of liquid pool and solid fuel fires can be estimated using empirical correlations or, in some cases, CFD computations.

Mixtures of fuel gases and air are flammable when the fuel concentration is between the lower and upper flammability limits. In a dynamic fuel release event, it is reasonable to expect that a flammable mixture will be found in some part of the volume. To ignite a flammable mixture of fuel and air, a local source of minimum ignition energy is needed. For most hydrocarbons, the minimum ignition energy is about 0.2 mJ. It is clear that the mechanical impacts of large vehicles and aircraft provide sufficient energy to ignite a flammable gas mixture. Above the auto-ignition temperature, which for most hydrocarbons ranges between 250 and 650°C, the gas mixture will ignite without an external source of energy.

Liquid fuels are classified according to their 'flash point', the temperature at which the vapour mixture above the liquid surface can be ignited. The flash points of jet fuels are higher than 38°C and, usually, above 60°C (see chapters 2–15 of Ref. [22]).

The ignition of solid fuels is a complicated dynamic process for which it is difficult to create simple rules that would establish whether the material would or would not ignite. Under continuous heating, many solid materials cannot be ignited if the radiant heat flux is less than about 11 kW/m^2, but this cannot be used as a rule in dynamic fire safety assessment. The thermal decomposition temperatures of solid materials typically range between 250 and 400°C. Once ignited, the flames will spread on the surface of the material. The flame spread rate depends on the material, geometry and ambient conditions. Empirical correlations have been derived for flame spread (appendix R of Ref. [25]) and peak heat release rate of cable tray fires [26]. Consideration of the ambient conditions is possible with CFD analysis using pyrolysis models to describe the heat transfer and degradation reactions of the solid.

2.4.3. External fire analysis

This section presents methods to assess external fire conditions. The goal of the assessment is to calculate the thermal exposure to external plant structures and SSCs. The thermal exposure is quantified in terms of radiative and convective heat flux incident on the target surface and the duration of the exposure. Sections 2.4.3.1 and 2.4.3.2 provide methods to assess external fireballs and pool fires, respectively. Section 2.4.3.3 provides assessment methods for aircraft impact fires.

2.4.3.1. Fireball

Fireballs can result from a sudden release and ignition of combustible liquid or gas. Such an event can be caused by an aircraft impact or by an explosive rupture of a pressurized fuel tank. Road and rolling-stock tankers can contain both gaseous and liquid hydrocarbons, including liquefied petroleum gas. In collision and fire situations, the rupture of pressurized gas tanks leads to a very rapid decrease in tank pressure and boiling of the fuel. The resulting two phase mixture is blown out, forming a fuel-rich droplet cloud which typically results in a spectacular fireball. This phenomenon is generally known as boiling liquid expanding vapour explosion (BLEVE). In fire situations, the tank rupture is caused by the loss of strength in the heated part of the tank. A review of the mechanisms, consequence assessment and management of BLEVEs is presented in Ref. [27].

From experiments and theoretical analyses, it can be concluded that a BLEVE fireball consists of three phases:

(a) The initial phase, in which a liquid–vapour mixture is blown out, forming a combustible mixture cloud;
(b) The expansion phase, in which the burning and expanding cloud forms a hemispherical fireball that sticks to the ground;
(c) The uplift phase, in which the hot fireball starts to rise in a more or less spherical mushroom-shaped plume.

The main parameter controlling the fireball properties is the mass of fuel involved in combustion. A conservative estimate is to assume that all of the fuel contributes to the fireball. For pressurized fuel tanks where the fuel temperature is only slightly above the boiling temperature, a more accurate estimate can be obtained using the concept of adiabatic flash, which relates the evaporated mass fraction to the ratio of sensible enthalpy stored in a heated liquid and the latent heat required for evaporation:

$$X_{AF} = \frac{c_p(T_T - T_B)}{H_v} \tag{18}$$

where

c_p is the liquid specific heat capacity (J/kg·K);
T_T is the liquid temperature (K);
T_B is the saturation (boiling) temperature (K);

and H_v is the liquid heat of vaporization (J/kg).

Empirical observations indicate that the actual mass contributing to the fireball during a BLEVE is higher than the adiabatic flash fraction X_{AF}. The following rule has been proposed for the fireball mass fraction X_{FB} [18]:

$$X_{FB} = \begin{cases} 1, \text{ if } X_{AF} > 1/3 \\ 3X_{AF}, \text{ if } X_{AF} < 1/3 \end{cases} \tag{19}$$

The mass of fuel in the fireball (in kilograms) is then calculated as:

$$M = X_{FB}M_T \tag{20}$$

where M_T is the total amount of fuel (kg).

In an aircraft impact, the amount of jet fuel burning in a fireball depends on the amount of fuel that penetrates into the target building and the amount that ends up burning in the pool fires below the impact location.

A simple formula has been derived for the maximum diameter D (in metres) of a hydrocarbon fireball by assuming that the fireball is formed by isochoric combustion followed by isothermal expansion.

$$D = 5.8M^{1/3} \tag{21}$$

The duration of a fireball t_d (in seconds) is harder to determine than the maximum diameter, since fireball evolution consists of the three phases mentioned above. From scaling arguments, it has been concluded that the fireball duration is proportional to the power of one third of the fuel mass if the fireball growth is dominated by expansion (momentum), and to the power of one sixth of the fuel mass if it is dominated by buoyancy. From observations, it has been concluded that large fireballs are buoyancy dominated. The following formula is consistent with these observations and can be used to calculate the fireball duration [18]:

$$t_d = \begin{cases} 0.45M^{1/3}, \text{ if } M < 30\ 000 \text{ kg} \\ 2.6M^{1/6}, \text{ if } M \geq 30\ 000 \text{ kg} \end{cases} \tag{22}$$

As Eqs (21, 22) reflect the averages of many other formulas for D and t_d, their use in the calculation of the final diameter and duration of BLEVE fireballs is recommended [20]. The height to which the fireball centre rises ranges from $0.75D$ to $1.33D$ (measured from the source of the fuel).

In spite of the relatively short duration, the thermal radiation of a fireball may cause burns to unprotected skin and ignite combustible materials. In tests with 100 kg to 100 t of liquid fuel, the values for emissive power E ranged from 80 to 200 kW/m^2; and from 150 to 330 kW/m^2 for kerosene and gasoline, respectively [28]. The average E of the fireball measured in tests of 1000 and 2000 kg butane and propane fireballs ranged from 320 to 370 kW/m^2. Based on these tests, a value of 350 kW/m^2 is recommended for surface emissive power E of BLEVE fireballs involving 1000 kg or more of vapour [20]. This value corresponds to radiation from a black body at a temperature of 1300°C.

The incident radiative heat flux $\dot{q}''_{r,in}$ (in watts per square metre) at a surface outside the fireball is usually calculated by modelling the fireball with a solid sphere emitting thermal radiation with a surface emissive power of E:

$$\dot{q}''_{r,in} = \tau F \cdot E \tag{23}$$

where

τ is atmospheric transmissivity;

and F is the geometric view factor.

For a point on a plane surface at distance r (in metres) from the fireball centre, the view factor is:

$$F = \left(\frac{D}{2r}\right)^2 \cos\theta \tag{24}$$

where θ is the angle between the normal of the surface and a line connecting the point to the fireball centre [20]. The atmospheric transmissivity τ may be calculated from the formula:

$$\tau = 2.02(p_w r)^{-0.09} \tag{25}$$

where

p_w is the partial pressure of water vapour in air (Pa);

and r is the path length (m) [29].

In summary, the consequences to a specified target due to a fireball can be calculated by:

— Estimating the mass of fuel M in the fireball;
— Calculating the fireball diameter, using Eq. (21);
— Calculating the fireball duration, using Eq. (22);
— Calculating/assuming the emissive power of thermal radiation ($E = 350$ kW m^2);
— Calculating the incident thermal radiation heat flux to the target, using Eq. (23);
— Calculating the thermal and mechanical response (Section 5.5).

The above procedure assumes that the convective heat flux from the fireball is small in comparison to the radiative heat flux and can be neglected in the analysis. It is important to recognize that BLEVEs can also cause flying missiles capable of penetrating structures and causing damage far away from the original location of the event [27].

2.4.3.2. Pool fire

The consequence to a specified target due to a pool fire can be calculated by:

— Estimating the mass M of the fuel in the pool;
— Calculating the fuel mass loss rate $\dot{M}''$ considering the radiation effects, using Eq. (26);
— Calculating the duration t_p of the pool fire, using Eq. (27);
— Calculating the flame length H_f, using Eq. (28);
— Calculating the flame surface emissive power E, using Eq. (29);
— Calculating the incident thermal radiative heat flux to the target surface, using Eq. (30);
— Calculating the total heat flux as a sum of convective and radiative fluxes, including a safety factor;
— Calculating the temperature and stress distributions through the structure (Section 5.5).

A pool fire burns at a constant rate, usually expressed as pool mass loss rate $\dot{M}''$ (kg $\cdot$ m^{-2} $\cdot$ s^{-1}):

$$\dot{M}'' = \dot{M}''_{\infty}(1 - e^{-k\beta D}) \tag{26}$$

where

the product $k\beta$ is represented as a single value (m^{-1}) given by k (m^{-1}), which is termed 'extinction coefficient', and β, the mean beam length correction;

and the quantity $\dot{M}''_{\infty}$ is the mass loss rate of an infinite pool and D is the effective pool diameter (m).

Reference [30] gives the following values for kerosene: $\dot{M}''_{\infty} = 0.039$ kg $\cdot$ m^{-2} $\cdot$ s^{-1} and $k\beta = 3.5$ m^{-1}; and for jet fuel type JP5: $\dot{M}''_{\infty} = 0.054$ kg $\cdot$ m^{-2} $\cdot$ s^{-1} and $k\beta = 1.6$ m^{-1}. Reference [31] gives a common set of values for kerosene and

JP5: $\dot{M}''_\infty = 0.063$ kg · m^{-2} · s^{-1} and $k\beta = 1.296$ m^{-1}, which presumably produces conservative predictions for kerosene.

Pool mass loss rate can be used to estimate the duration of pool fires. If the mass loss rate is constant, the pool fire duration is calculated as:

$$t_p = \frac{M}{\dot{M}''A} \tag{27}$$

where A is the pool area. Alternatively, dividing the mass loss rate by the liquid density gives the liquid regression rate (in metres per second). The estimated pool fire duration is then the pool depth divided by the liquid regression rate. If the liquid is spilled at a constant rate, an estimate for liquid pool area can be found by dividing liquid spill rate by liquid regression rate.

For calculations of the thermal radiation effects, pool fires are usually represented by a solid cylinder whose dimensions correspond to the time-averaged dimensions of the flames, as illustrated in Fig. 12. Wind tilts the flames and causes the flame base to extend beyond the downwind pool edge. This effect is

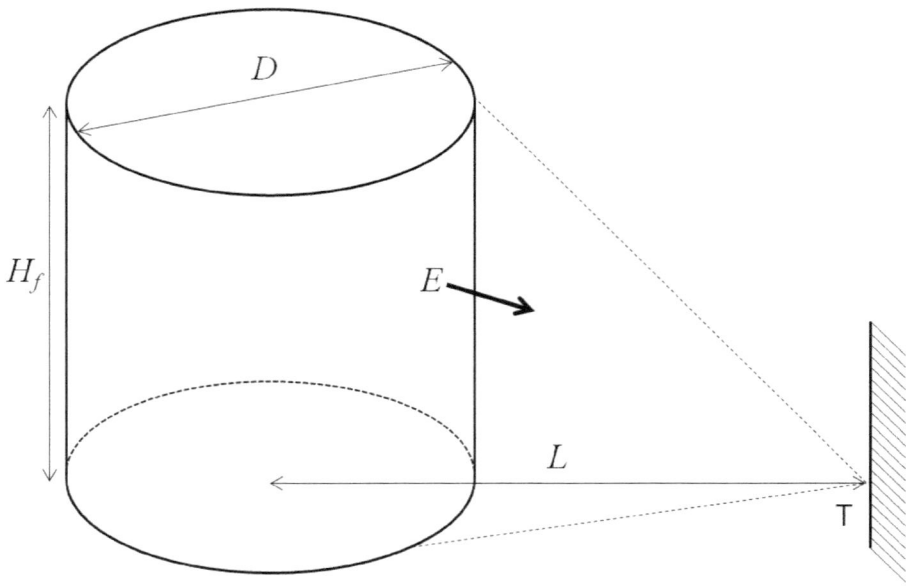

FIG. 12. Modelling the pool fire flame as a cylinder for radiation calculation. The heat flux $\dot{q}''_{r,in}$ is calculated at target T.

called flame drag. Empirical correlations for flame behaviour are summarized in Ref. [32].

The flame length H_f (in metres) of circular or close-to-circular pool fires can be calculated using the Heskestad correlation:

$$H_f = 0.235\dot{Q}^{2/5} - 1.02D \tag{28}$$

where $\dot{Q}$ is the fire heat release rate (kW).

The average surface emissive power E of kerosene pool fires decreases with increasing pool diameter D according to the empirical correlation for this quantity:

$$E = E_f e^{-sD} + E_s(1 - e^{-sD}) \tag{29}$$

where E_f and E_s are the surface emissive power of flame (140 kW/m²) and smoke (20 kW/m²), respectively. These emissive powers correspond to radiation from black bodies at temperatures of 980 and 500°C. The parameter s has a value of 0.12 m⁻¹ [29, 32].

The incident radiative heat flux to a target surface is:

$$\dot{q}''_{r,in} = \tau F \cdot E \tag{30}$$

where

F is the view factor between the flame and the target surface

and τ is the atmospheric transmissivity (Eq. (24)).

Formulas for the view factor of vertical and tilted cylinders are given in Refs [29, 32]. These formulas can be applied to targets on the pool level, which is usually the ground, as illustrated in Fig. 12. The view factor from a vertical cylinder of diameter D and height H_f to a vertical surface at distance L from the cylinder centre is:

$$F = \frac{1}{\pi S} \tan^{-1}\left(\frac{h}{\sqrt{S^2 - 1}}\right) - \frac{h}{\pi S} \tan^{-1} \sqrt{\frac{S-1}{S+1}} +$$
$$\frac{A \cdot h}{\pi S \sqrt{A^2 - 1}} \tan^{-1} \sqrt{\frac{(A+1)(S-1)}{(A-1)(S+1)}} \tag{31}$$

where

$$A = \frac{h^2 + S^2 + 1}{2S};$$

$$S = \frac{2L}{D};$$

$$h = \frac{2H_f}{D}.$$

For elevated targets, the cylinder is divided into two parts and the view factor is the sum of the respective view factors of these parts.

The total net heat flux to the target surface is calculated as the sum of the convective and radiative heat fluxes. The convective heat flux is significant when the target is inside the flame. The calculation of the convective heat flux can be based on the maximum average gas temperature of the turbulent buoyant flame, which is generally observed to be in the range of 900 to 1100°C, and the convective heat transfer coefficient $h_c = 0.02 \ \text{kW} \cdot \text{m}^{-2} \cdot \text{K}$.

Application of the above correlations for the calculation of the flame length and heat flux for a pool fire resulting from an aircraft impact is problematic because the shape of the pool may not be even close to a circle. An alternative set of correlations has been derived for trench fires (summarized in Ref. [32]). For wider generality, the use of CFD computations is to be considered. When using the above correlations for design purposes, a safety factor of two is to be applied on the predicted heat flux levels. The relatively large safety factor is caused by the large scatter within the experimental data and various empirical correlations.

2.4.3.3. Aircraft crash

The thermal effects from an aircraft crash outside the plant buildings can be evaluated using hand calculation methods or CFD analysis. When using hand calculations, the amount of fuel is first divided between the fireball and the pool fire. Afterwards, both phenomena are evaluated separately using the methods described in Sections 2.4.3.1 and 2.4.3.2.

The fraction of the aircraft fuel that burns in a fireball strongly depends on the vertical distance of the impact location to the horizontal surface potentially ponding the fuel (e.g. ground, auxiliary building roof) [33]. For small distances, about 50% of the fuel burns in the fireball. For high impact locations, the majority of the fuel will burn in the fireball. The area of the pool depends on the geometry and impact conditions. A rough estimate can be obtained by assuming that the fuel is collected as a layer of 30 kg/m^2 [24]. The resulting pool areas for the aircraft size groups of Table 3 are shown in Table 4, assuming a 50% pooling fraction.

TABLE 4. SUGGESTED POOL SIZES RESULTING
FROM AN AIRCRAFT CRASH

Aircraft type	Group A	Group B	Group C
Pool area (m^2)	300	900	3000

The duration of the pool fire can be calculated from the estimated pool size, mass and burning rate. The firefighting measures of a fire brigade, if successful, may reduce the calculated fire duration significantly. However, it is very unlikely that effective firefighting measures could be performed for pools as large as those listed in Table 4.

Documentary footage of aircraft crashes shows that most aircraft crashes, whether they occur on hard runways or soft terrain, result in an immediate and large fireball. These fireballs show all the features of fireballs from ruptured tanks containing pressure liquefied hydrocarbon gases [34], meaning that the methods described in Section 2.4.3.1 will be well suited for the analysis. The effects of the hard vertical target structure, geometrical details and wind conditions will, however, require the use of CFD simulations. CFD simulations are also needed if the interaction between the fire and plant structures needs to be analysed in detail. Similar constraints are related to the use of fire plume models in assessing the consequences of pool fires. Fire plume models are valid for close-to-circular pools burning on flat ground in the open atmosphere, and still or moderate wind conditions. Complex geometries and pool shapes need to be studied with CFD analysis.

From the fluid dynamics viewpoint, aircraft impact fires are very different from compartment fires, for which the fire CFD models are usually applied. The applicability of a specific CFD code to the problem can be evaluated from the following requirements:

— The CFD code needs to be able to handle highly dynamic reactive flows. Turbulence modelling is to be based on large eddy simulation rather than Reynolds-averaged Navier–Stokes.
— If using a two phase formulation, the code needs to allow for describing the spatial and temporal characteristics of the initial spray. The shape of the droplet size distribution can be either prescribed, if the necessary

experimental data are available, or predicted using appropriate models for droplet formation. Simulation of the primary atomization process is not considered possible for this kind of complicated release event, but the models of secondary break-up can produce physically meaningful distributions once the initial drop size has been prescribed.

— Droplet movement and drag between the droplets and gas need to be calculated accurately enough to capture the dynamic spreading, air entrainment and droplet cloud dispersion. The high relative velocity between the droplets and the gas requires good temporal resolution and accurate conservation of momentum.
— The code needs to include a thermal radiation model.
— The convective and radiative heating of the droplets is to be taken into account. The absorption and scattering coefficients of the spray can be calculated using Mie theory and appropriate radiative properties. As these properties may not be available for the jet fuel, some generic properties for hydrocarbon liquid fuels can be used.
— The evaporation of the fast moving droplet needs to account for the film mass transfer and high temperature differences.
— The convective and radiative heat transfer from the combustion cloud needs to be simulated. It is reasonable to assume that the evaporated fuel will ignite immediately. An assumption of fast chemistry can be made but its implications on the simulation results are currently not known.

The most important boundary condition of the CFD fire analysis is the fire source description. Two different methods can be used for introducing the fuel into the computational domain. The first method is to replace the source of liquid jet fuel as an equivalent source of gaseous fuel. This method can be used to simulate the overall fireball and fire plume behaviour [34]. The release time should correspond to the duration of the impact process. This method cannot capture the effects of the fuel spray dynamics affecting both the distribution of the fuel between the fireball and the pool fire, and the expansion of the fireball.

An alternative, more realistic method of introducing the fuel into the calculation describes the jet fuel using a two phase flow calculation. Such calculations have, up to this point, been performed using the fire dynamics simulator code [33, 35]. The following procedure can be used as a guideline for setting up such simulations.

Geometrical shape of the spray

The fuel spray is released from the ruptured fuel tanks of the aircraft as a result of the initial kinetic energy and pressure of the crashing aircraft structures.

The initial shape of the release can be very complicated because modern aircraft contain several tanks in the wings and hull of the plane. A general observation from experiments with water-containing aircraft or missiles crashing into a vertical wall is, however, that the water spreads as a thin sheet along the direction of the wall in a relatively symmetrical pattern. Analysis of the Sandia Phantom F-4 experiment [36] shows that the initial spray cloud can be approximated by a ring that is close to the impact wall surface and surrounds the aircraft impact location. The radius of the ring is to be chosen to expand on multiple grid cells. The direction of the droplets in the Phantom experiment ranged between 0° and 30°±10° from the wall tangent. No liquid was released in the direction of the wings. In the tests with cylindrical steel missiles [37], the droplets were also released at an angle between 0° and 30°. Further away from the initial release point, the maximum angle of the spray was only 10° to 15° from the wall tangent due to the one-sided entrainment of air into the spray.

A schematic picture of the liquid release pattern for the simulation boundary condition is shown in Fig. 13. The droplets are introduced onto a spherical

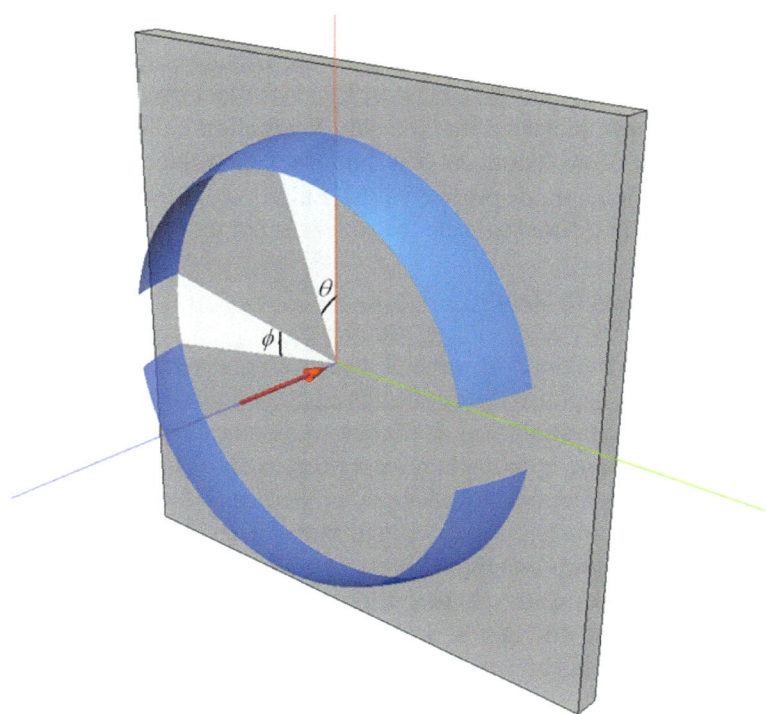

FIG. 13. *Virtual surface for introducing droplets into the calculation domain.*

surface at distance R from the impact centre location and with angles $\theta = 30°$ and $\varphi = 15°$. Distance R is to be greater than the aircraft fuselage radius. In addition, the sphere diameter $2R$ needs to span about ten grid cells for sufficient resolution of the initial spray.

Initial speed of the spray

An estimate for the initial liquid spray speed in the Sandia F-4 experiment is 1.3 to 1.55 times the impact speed. The uncertainty of this estimate is quite high, and the estimates could not be made sufficiently close to the actual water release location. Experiments with water filled cylinders indicate that the initial speed of the spray is 2 to 2.5 times the impact speed [37].

The recommended value for the initial droplet speed is twice the impact speed of the aircraft.

Initial droplet size distribution

Owing to the difficulties associated with using numerical tools to predict the droplet size distribution, it is recommended that the size distribution be determined experimentally. The experiments are usually performed with water for safety reasons. Experiments with water filled steel cylinders impacting a rigid wall at 100 m/s have indicated that the size distribution is bimodal, and can be modelled as a sum of two unimodal distributions with arithmetic mean diameters of around 0.2 and 0.6 mm, respectively [38]. Empirical correlations for the mean droplet diameter can be used to scale the water results for jet fuel [39].

Thermal properties of the liquid fuel

Jet fuel is a mixture of several hydrocarbons and difficult to model accurately for fire chemistry. For the computation of droplet heat transfer, evaporation and combustion, it is sufficient to assume that the droplets consist of some relatively large hydrocarbon molecules, such as heptane C_7H_{16}, and to apply the appropriate thermal and chemical properties accordingly.

The numerical models for the above phenomena need to be verified and validated. The validation conditions are to be similar to the conditions of the aircraft crash application. In practice, the validation needs to be performed in several stages for the different aspects of an aircraft crash fire because well documented experiments of a complete event do not exist.

2.4.4. Internal fire analysis

The fire conditions inside the buildings and the consequent load on the safety related structures and components are evaluated using fire safety engineering calculations and fire simulations with specific computer codes. The most commonly used fire safety engineering calculation methods have been developed for room fires. They usually assume that the fire takes place in a closed or naturally ventilated compartment with door-like ventilation to the open atmosphere. These calculation methods are of limited value in the assessment of fireball-initiated, rapidly developing fires but may be used, in some instances, for analysing conventional fires following an external event. A collection of fire safety engineering tools for internal fire analysis can be found in, for example, Ref. [40].

CFD methods are generally valid for rapidly developing fires. Using CFD requires more detailed specification of the geometry and more computational resources. The modelling guidelines for CFD codes are discussed in Ref. [41].

The fire heat release rate is controlled either by the area of the burning fuel or by the available oxygen. It is typical to assume that the fire starts as a fuel controlled fire, increases in size and turns into a ventilation controlled fire. In very large spaces, the fire can remain fuel controlled over the entire duration of the fire event. These assumptions do not hold for fires that result from instantaneous ignition of large surfaces, such as when a fireball enters a building after an event. In such cases, the fire will become ventilation controlled within a few seconds. The ventilation conditions are, therefore, to be estimated according to the damage incurred after a human induced event. The mechanical ventilation and filter systems may be out of operation due to physical or electrical effects, which can be easily verified with respect to the human induced event scenario.

A plant walkdown can be performed to obtain vital information for computer fire simulations. Plant walkdowns need to be performed by multidisciplinary teams of knowledgeable people, i.e. fire engineering specialists are to be accompanied by engineering specialists of civil/structural, mechanical, instrumentation and control, electrical and systems disciplines. The information gathered needs to include the location and type of critical equipment and cable trays, the separation between redundant trains and shielding, and the fire barriers and fire extinguishing capabilities that may be present.

3. MATERIAL PROPERTIES

3.1. GENERAL CONSIDERATIONS

Numerical simulation of an impact loaded, reinforced concrete structure is a challenging task, since the material is inhomogeneous and the material behaviour is non-linear and strain rate and temperature dependent.

Depending on the characteristics of the impact load and on the geometry of the impacted structure, the dominating failure mode can be local or global. Local failure modes due to an impact are perforation and scabbing. Global failure modes are mainly due to bending moments and axial and shear forces affecting the structural members between the impact location and the support points. These global failure modes may cause structures to collapse.

The behaviour of concrete in compression and its compressive strength are important in assessing the local behaviour of concrete structures subjected to an impact with hard missiles. Tensile strength properties and the stress–strain response of reinforcement steel, in many cases, dominate the deflection and vibration behaviour of the reinforced concrete structural members. In analysing the ultimate capacity of a structure, the use of realistic material properties is justified. The variation of material properties with age also needs to be taken into account in the assessment.

3.2. CONCRETE

3.2.1. Concrete material properties

The main challenge in non-linear analyses of concrete structures is the relatively low tensile strength of concrete. The amount of tensile stress causing cracking in concrete is on the order of one tenth of the crushing strength of the concrete. In reinforced concrete structures, the low tensile strength is compensated for by reinforcement.

A typical one dimensional stress–strain response of concrete is shown schematically in Fig. 14. A typical biaxial stress state failure curve for plain concrete is presented in Fig. 15. In this figure, σ_I and σ_{II} are the principal stresses and β_p is the uniaxial compressive strength of concrete. The angle α can be defined using the principal stresses as $\tan\alpha = \sigma_I / \sigma_{II}$.

Biaxial stress states can be classified as follows:

— $0 \leq \alpha \leq \dfrac{\pi}{2}$: biaxial tension stress state.

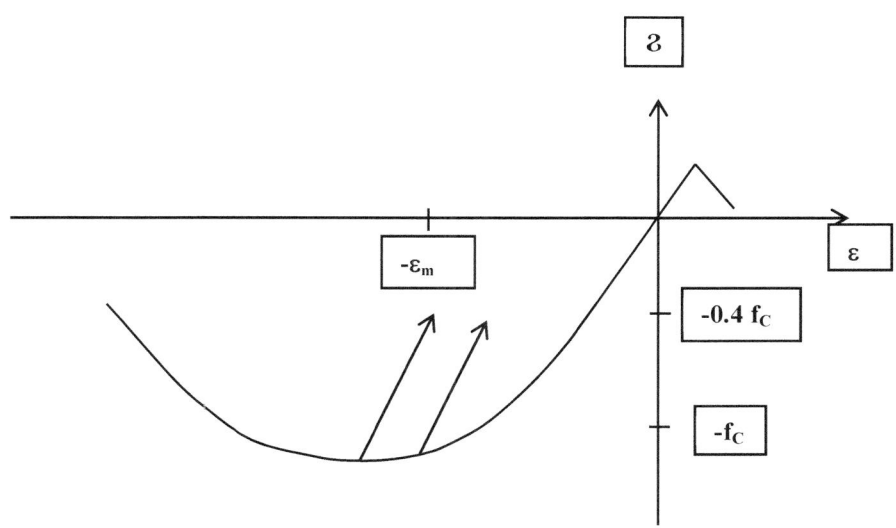

FIG. 14. One dimensional stress–strain behaviour of concrete.

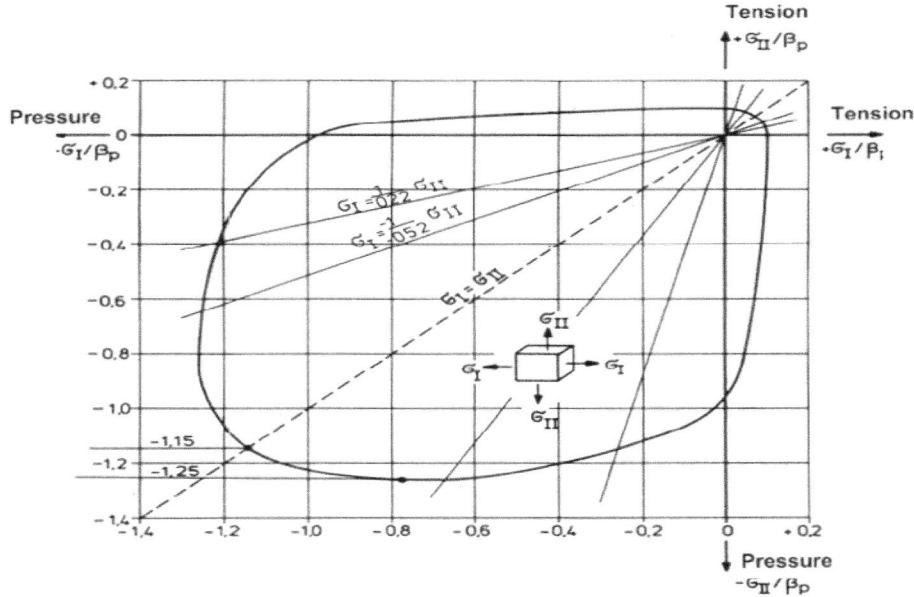

FIG. 15. Biaxial stress state of concrete.

49

— $\frac{\pi}{2} < \alpha < \pi$ and $\frac{3\pi}{2} < \alpha < 2\pi$: combined tension–compression stress state.

— $\pi \leq \alpha \leq \frac{3\pi}{2}$: biaxial compression stress state.

The tensile strength of concrete is almost independent of the ratio of principal stresses. The biaxial compressive strength of concrete is larger than the uniaxial compressive strength.

A typical triaxial stress state failure surface for plain concrete is presented in Fig. 16. In this figure, each stress state is represented by a point $(\sigma_I, \sigma_{II}, \sigma_{III})$ in an orthogonal coordinate system, where σ_I, σ_{II} and σ_{III} are the principal stresses. As shown in the figure, σ_o and τ_o are the octahedral normal and shear stresses in the π plane; σ_o is the hydrostatic stress; and τ_o the deviatoric stress. The direction of τ_o in the deviatoric plane is defined by the angle θ ($\theta = 0°$: $\sigma_I = \sigma_{II} < \sigma_{III}$; $\theta = 60°$: $\sigma_I < \sigma_{II} = \sigma_{III}$). At the failure surface, when $\theta = 0°$, $\tau_{oc} = \tau_o$; and when $\theta = 60°$, $\tau_{oe} = \tau_o$.

Under compression, concrete behaves rather linearly elastic until roughly 40% of the ultimate compression strength is reached. Beyond this point, non-linear behaviour starts. Unloading after reaching the inelastic range causes permanent deformations. Young's modulus of concrete increases somewhat with stress level and with strain rate under compressive stresses.

3.2.2. Concrete material properties at elevated strain rates

Concrete material strength increases at elevated strain rates. The dynamic strength values may be applied in structural analyses when properly justified. The

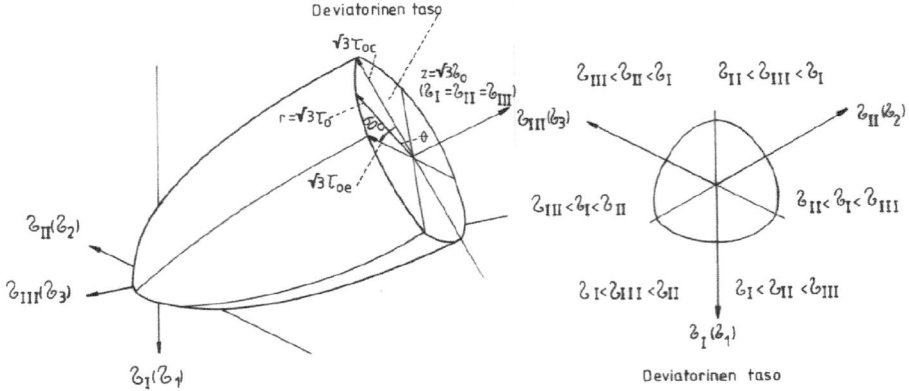

FIG. 16. Triaxial stress state of concrete and π plane presentation.

strength of concrete at elevated strain rates is higher than the material strength values obtained by static material tests. The dynamic increase factor, that is the ratio of the dynamic to static strength, is normally reported as a function of strain rate. These dynamic increase factors are applicable to simplified models, as in Ref. [42].

Regarding finite element codes, modelling techniques for strain rate dependent material properties are code specific. For concrete, the dominant source of dynamic increase of strength is concrete confinement. This confinement is a consequence of inertial effects, which are inherent to finite element modelling. Care should be taken so that dynamic increase factors are not considered twice when using finite element models.

3.2.2.1. Compressive strength at elevated strain rates

Experimental results on compressive behaviour at high strain rates are reviewed and reported, for example, in Ref. [43]. A compilation of test results on relative compressive strength increase is shown in Fig. 17. Up to a strain rate

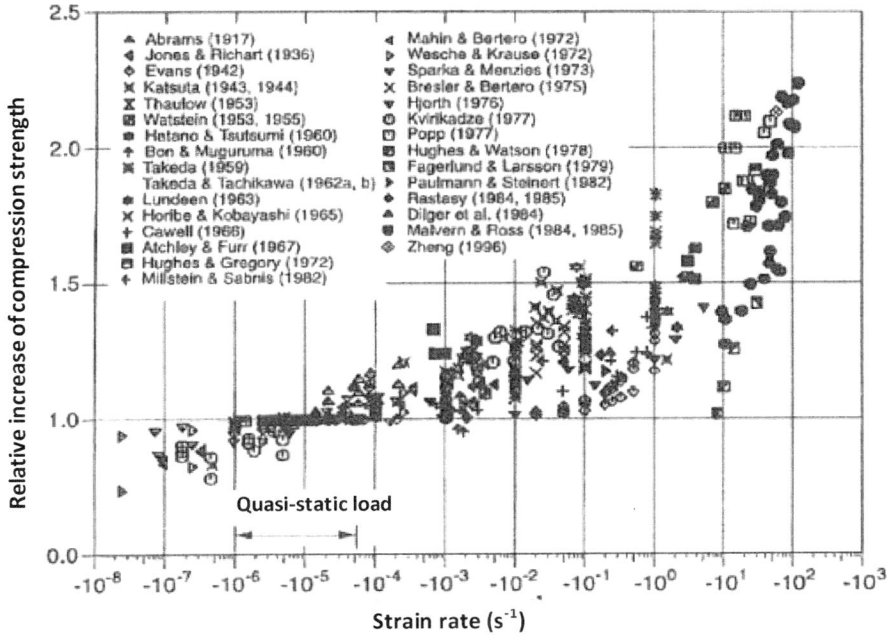

FIG. 17. Strain rate effects on unconfined compressive strength of concrete [43].

of 10 s^{-1}, there is not much deviation from static compressive strength. During impact loading at about a strain rate of 100 s^{-1}, the compressive strength of concrete has been found to be 100% higher than the static strength. However, it should be noted that there is a wide scatter in test results. The compressive strength of concrete is not as highly dependent on the strain rate as is the tensile strength (Section 3.2.2.2).

For the increase in peak compressive strength f'_c, a dynamic increase factor is introduced in the Comité euro-international du béton (Euro-International Concrete Committee)–Fédération internationale de la précontrainte (International Federation of Prestressing) model for strain rate dependence of concrete strength as follows:

$$\text{DIF} = \begin{cases} \left(\dfrac{\dot{\varepsilon}}{\dot{\varepsilon}_s}\right)^{1.026\alpha} & \text{for } |\dot{\varepsilon}| \leq 30 \text{ s}^{-1} \\[2ex] \gamma\left(\dfrac{\dot{\varepsilon}}{\dot{\varepsilon}_s}\right)^{\frac{1}{3}} & \text{for } |\dot{\varepsilon}| > 30 \text{ s}^{-1} \end{cases} \tag{32}$$

where

$\dot{\varepsilon}$ is the strain rate;
$\dot{\varepsilon}_s$ is 30×10^{-6} s^{-1} (quasi-static strain rate);
$\lg \gamma$ is $6.156\alpha - 2$;

and $\alpha = \dfrac{1}{5 + 9(f'_c / f_{c0})}$ with $f_{c0} = 10$ MPa.

Other formulations for the dynamic increase factor are also used in practice [9, 42, 44].

3.2.2.2. Tensile strength at elevated strain rates

Tensile strength increases at high strain rates. Figure 18 shows experimental data reported in the literature.

The tensile dynamic increase factor is obtained from:

$$\text{TDIF} = \begin{cases} \left(\dfrac{\dot{\varepsilon}}{\dot{\varepsilon}_s}\right)^{1.016\delta} & \text{for } |\dot{\varepsilon}| \leq 30 \text{ s}^{-1} \\[2ex] \beta\left(\dfrac{\dot{\varepsilon}}{\dot{\varepsilon}_s}\right)^{\frac{1}{3}} & \text{for } |\dot{\varepsilon}| > 30 \text{ s}^{-1} \end{cases} \tag{33}$$

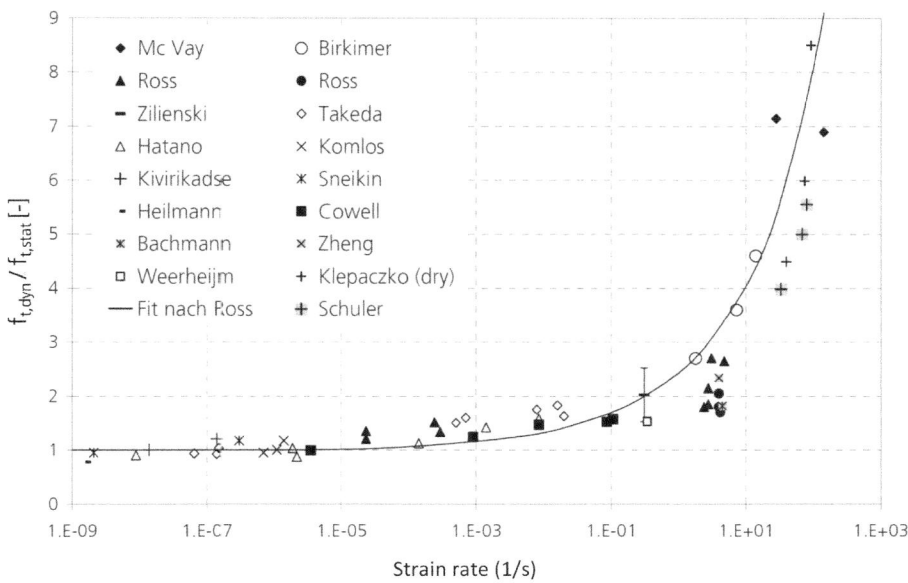

FIG. 18. Relative tensile strength increase for different loading rates [45].

with $\lg \beta = 7.112\delta - 2.33$

and $\delta = \left(10 + \dfrac{3}{5} f_{cm}\right)^{-1}$

where

$\dot{\varepsilon}$ is the tensile strain rate in the range of 3×10^{-6} to 300 s^{-1};

$\dot{\varepsilon}_s$ is 3×10^{-6} s^{-1};

and f_{cm} is the static uniaxial compressive strength (MPa).

The factor TDIF = 1.0 for strain rates lower than 3×10^{-6} s^{-1}. The strain rate is calculated from the average volumetric strain rate in the analyses.

3.2.3.   Concrete material properties at elevated temperatures

The compressive strength of concrete starts to decrease at a temperature range of 100 to 350°C, depending on the stress state. The compressive strength decreases remarkably at a temperature range of 400 to 600°C. Young's modulus decreases as the temperature increases. In addition, the bond between rebar and concrete weakens as the temperature increases.

3.2.3.1. Compressive strength at elevated temperatures

According to Ref. [46], the strength and deformation properties of uniaxial stressed concrete at elevated temperatures are obtained using a stress–strain relationship such as:

$$\sigma(\theta) = \frac{3\varepsilon f_{c,\theta}}{\varepsilon_{c1,\theta}\left[2 + \left(\dfrac{\varepsilon}{\varepsilon_{c1,\theta}}\right)^3\right]} \qquad \varepsilon \leq \varepsilon_{c1,\theta} \tag{34}$$

where

ε is the strain value;

$f_{c,\theta}$ is the compressive strength at temperature θ;

and $\varepsilon_{c1,\theta}$ is the corresponding compressive strain.

For the descending branch in the range of $\varepsilon_{c1(\theta)} < \varepsilon \leq \varepsilon_{cu1,\theta}$, linear and non-linear models are permitted. The stress–strain relationship is depicted in Fig. 19.

Values for these parameters are given in Table 5 as a function of concrete temperatures. These parameters may be used for normal weight concrete with siliceous or calcareous (containing at least 80% calcareous aggregate by weight) aggregates. It should be noted that a possible strength gain of the concrete in the cooling phase is not to be taken into account.

3.2.3.2. Tensile strength at elevated temperatures

According to Ref. [46], a reduction in the characteristic tensile strength of concrete is represented by the coefficient $k_{c,t}(\theta)$:

$$f_{ck,t}(\theta) = k_{c,t}(\theta) f_{ck,t} \tag{35}$$

where

$f_{ck,t}$ is the characteristic tensile strength of concrete;

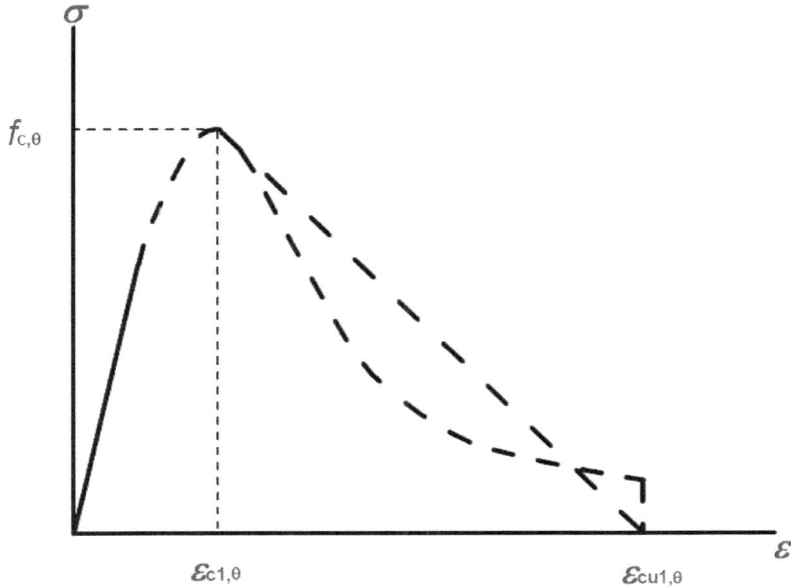

FIG. 19. Mathematical model for stress–strain relationships of concrete under compression at elevated temperatures [46] (©CEN, reproduced with permission).

and $f_{ck,t}(\theta)$ is the reduced tensile strength of concrete at temperature θ.

In the absence of more accurate information, the following $k_{c,t}(\theta)$ values can be used:

$$k_{c,t}(\theta) = 1.0 \qquad\qquad \text{for } 20°C \le 100°C$$

$$k_{c,t}(\theta) = 1.0 - 1.0(\theta - 100)/500 \qquad \text{for } 100°C \le \theta \le 600°C$$

3.3. STEEL

3.3.1. Steel material properties

Stress–strain curves for typical hot rolled and cold worked steel are presented in Fig. 20.

TABLE 5. VALUES FOR THE MAIN PARAMETERS OF THE STRESS–
STRAIN RELATIONSHIPS OF NORMAL WEIGHT CONCRETE WITH
SILICEOUS OR CALCAREOUS AGGREGATE CONCRETE AT ELEVATED
TEMPERATURES [46]*

Concrete temperature	Siliceous aggregates			Calcareous aggregates		
θ (°C)	$f_{c,\theta}/f_{ck}$	$\varepsilon_{c1,\theta}$	$\varepsilon_{cu1,\theta}$	$f_{c,\theta}/f_{ck}$	$\varepsilon_{c1,\theta}$	$\varepsilon_{cu1,\theta}$
20	1.00	0.0025	0.0200	1.00	0.0025	0.0200
100	1.00	0.0040	0.0225	1.00	0.0040	0.0225
200	0.95	0.0055	0.0250	0.97	0.0055	0.0250
300	0.85	0.0070	0.0275	0.91	0.0070	0.0275
400	0.75	0.0100	0.0300	0.85	0.0100	0.0300
500	0.60	0.0150	0.0325	0.74	0.0150	0.0325
600	0.45	0.0250	0.0350	0.60	0.0250	0.0350
700	0.30	0.0250	0.0375	0.43	0.0250	0.0375
800	0.15	0.0250	0.0400	0.27	0.0250	0.0400
900	0.08	0.0250	0.0425	0.15	0.0250	0.0425
1000	0.04	0.0250	0.0450	0.06	0.0250	0.0450
1100	0.01	0.0250	0.0475	0.02	0.0250	0.0475
1200	0.00	—[a]	—[a]	0.00	—[a]	—[a]

* ©CEN, reproduced with permission
[a] —: data not available.

3.3.2. Steel material properties at elevated strain rates

The yield strength of steel is highly strain rate dependent and increases
when the strain rate increases. The dynamic yield strength of steel can be taken

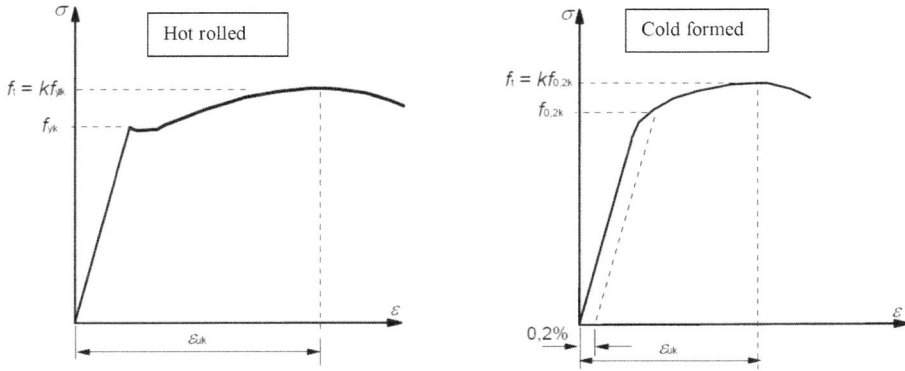

FIG. 20. Stress–strain diagrams of typical reinforcing steel.

into consideration by the Cowper–Symonds formula for uniaxial tension or compression:

$$\sigma_{yd} = \sigma_{ys}\left[1+\left(\frac{\dot{\varepsilon}}{D}\right)^{1/q}\right] \tag{36}$$

where

σ_{yd} and σ_{ys} are the static and the dynamic yield stress, respectively;

and D and q are material parameters; for mild steel, $D = 40\ \text{s}^{-1}$ and $q = 5$ can be used [47].

For pre-stressing steel, no dynamic yield strength increase is to be used.

3.3.3. Steel material properties at elevated temperatures

3.3.3.1. Reinforcement steel material properties at elevated temperatures

According to Ref. [46], the strength and deformation properties of reinforcing steel at elevated temperatures can be obtained from the following stress–strain relationships:

$$
\begin{cases}
\sigma(\theta)=\in E_{sp,\theta} \quad E_{\mathrm{T}}=E_{s,\theta} \quad \varepsilon < \varepsilon_{sp,\theta} \\[2mm]
\sigma(\theta)= f_{sp,\theta} - c + \left(\dfrac{b}{a}\right)\left[a^2 - \left(\varepsilon_{sy,\theta}-\varepsilon\right)^2\right]^{0.5} \quad E_{\mathrm{T}} = \dfrac{b\left(\varepsilon_{sy,\theta}-\varepsilon\right)}{a\left[a^2-\left(\varepsilon-\varepsilon_{sy,\theta}\right)^2\right]^{0.5}} \\[6mm]
\hspace{6cm} \varepsilon_{sp,\theta} \le \varepsilon \le \varepsilon_{sy,\theta} \\[2mm]
\sigma(\theta)= f_{sy,\theta} \quad E_{\mathrm{T}}=0 \quad \varepsilon_{sy,\theta} < \varepsilon < \varepsilon_{st,\theta} \\[2mm]
\sigma(\theta)= f_{sy,\theta}\left[1-\left(\varepsilon-\varepsilon_{st,\theta}\right)/\left(\varepsilon_{su,\theta}-\varepsilon_{st,\theta}\right)\right] \quad \varepsilon_{st,\theta} < \varepsilon < \varepsilon_{su,\theta} \\[2mm]
\sigma(\theta)=0 \quad \varepsilon = \varepsilon_{su,\theta}
\end{cases}
\tag{37}
$$

Further parameters are defined: $\quad \varepsilon_{sp,\theta} = \dfrac{f_{sp,\theta}}{E_{s,\theta}} \qquad \varepsilon_{sy,\theta} = 0.02$

For cold worked reinforcement: $\quad \varepsilon_{st,\theta} = 0.05 \qquad \varepsilon_{su,\theta} = 0.10$

For hot rolled reinforcement: $\qquad \varepsilon_{st,\theta} = 0.15 \qquad \varepsilon_{su,\theta} = 0.20$

The functions a, b and c are:

$$
a^2 = \left(\varepsilon_{sy,\theta} - \varepsilon_{sp,\theta}\right)\left(\varepsilon_{sy,\theta} - \varepsilon_{sp,\theta} + \dfrac{c}{E_{s,\theta}}\right)
$$

$$
b^2 = c\left(\varepsilon_{sy,\theta} - \varepsilon_{sp,\theta}\right)E_{s,\theta} + c^2
$$

$$
c = \dfrac{\left(f_{sy,\theta} - f_{sp,\theta}\right)^2}{\left(\varepsilon_{sy,\theta} - \varepsilon_{sp,\theta}\right)E_{s,\theta} - 2\left(f_{sy,\theta} - f_{sp,\theta}\right)}
$$

where

$E_{s,\theta}$ is the slope of the linear-elastic range;
$f_{sp,\theta}$ is the proportionality limit;
$\varepsilon_{sp,\theta}$ is the corresponding strain;
$f_{sy,\theta}$ is the maximum stress level;
$\varepsilon_{sy,\theta}$ is the corresponding strain;
$\varepsilon_{st,\theta}$ is the strain value after which the stress decreases as the strain increases;

and $\varepsilon_{su,\theta}$ is the ultimate strain corresponding to zero stress.

The stress–strain relationship and the associated parameters are illustrated in Fig. 21. Values for parameters of the stress–strain relationship of hot rolled and cold worked reinforcing steel at elevated temperatures are given in Table 6.

3.3.3.2. Pre-stressing steel material properties at elevated temperatures

According to Ref. [46], the strength and deformation properties of pre-stressing steel at elevated temperatures may be obtained by the same mathematical model as presented above for reinforcing steel. In Eq. (37), subscript 's' changes to 'p' for pre-stressing steel. Values for $\varepsilon_{pt,\theta}$ and $\varepsilon_{pu,\theta}$ for pre-stressing steel may be taken from table 3.3 of Ref. [46].

Values for the parameters of the stress–strain relationship for cold worked (wires and strands) and quenched and tempered (bars) pre-stressing steel at elevated temperatures are given in Table 7. For pre-stressing steel, the yield stress at room temperature f_{yk} that appears in Table 7 can usually be taken as 0.9 times the tensile strength f_{pk}.

3.3.3.3. Structural steel material properties at elevated temperatures

For the low-carbon steels typically used in building construction, the strength and deformation properties at elevated temperatures may be obtained by the same mathematical model as presented in Section 3.3.3.1 for hot rolled reinforcing steel [48].

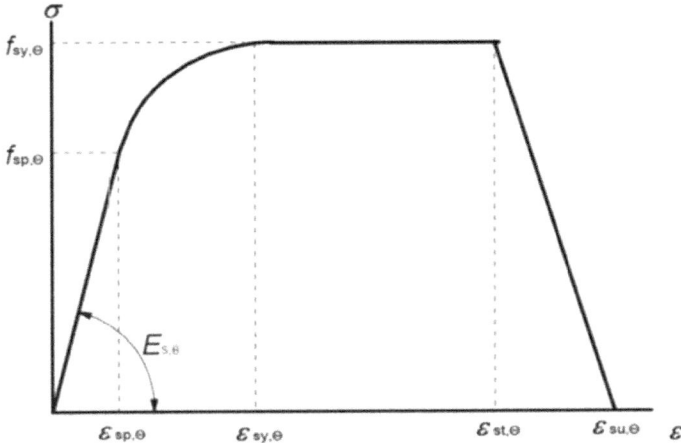

FIG. 21. Mathematical model for stress–strain relationships of reinforcement and pre-stressing steel at elevated temperatures [46] (© CEN, reproduced with permission).

TABLE 6. VALUES FOR THE PARAMETERS OF THE STRESS–STRAIN RELATIONSHIP OF HOT ROLLED AND COLD WORKED REINFORCING STEEL AT ELEVATED TEMPERATURES [46]*

Steel temperature	$f_{sy,\theta}/f_{yk}$		$f_{sp,\theta}/f_{yk}$		$E_{s,\theta}/E_p$	
θ (°C)	Hot rolled	Cold worked	Hot rolled	Cold worked	Hot rolled	Cold worked
20	1.00	1.00	1.00	1.00	1.00	1.00
100	1.00	1.00	1.00	0.96	1.00	1.00
200	1.00	1.00	0.81	0.92	0.90	0.87
300	1.00	1.00	0.61	0.81	0.80	0.72
400	1.00	0.94	0.42	0.63	0.70	0.56
500	0.78	0.67	0.36	0.44	0.60	0.40
600	0.47	0.40	0.18	0.26	0.31	0.24
700	0.23	0.12	0.07	0.08	0.13	0.08
800	0.11	0.11	0.05	0.06	0.09	0.06
900	0.06	0.08	0.04	0.05	0.07	0.05
1000	0.04	0.05	0.02	0.03	0.04	0.03
1100	0.02	0.03	0.01	0.02	0.02	0.02
1200	0.00	0.00	0.00	0.00	0.00	0.00

* ©CEN, reproduced with permission

Hence, values for the parameters of the stress–strain relationship for structural steel at elevated temperatures are given in columns 2, 4 and 6 of Table 6, which correspond to the 'hot rolled' material.

It should be noted that significant changes in crystalline structure begin to occur in this type of steel at metal temperatures in excess of 650°C. These changes lead to a sharp reduction of mechanical capacities above this temperature threshold [22].

TABLE 7. VALUES FOR THE PARAMETERS OF THE STRESS–STRAIN RELATIONSHIP OF COLD WORKED, AND QUENCHED AND TEMPERED PRE-STRESSING STEEL AT ELEVATED TEMPERATURES [46]*

Steel temperature	$f_{py,\theta}/f_{yk}$		$f_{pp,\theta}/f_{yk}$		$E_{p,\theta}/E_p$	
θ (°C)	Quenched and tempered	Cold worked	Quenched and tempered	Cold worked	Quenched and tempered	Cold worked
20	1.00	1.00	1.00	1.00	1.00	1.00
100	0.98	0.99	0.77	0.68	0.76	0.98
200	0.92	0.87	0.62	0.51	0.61	0.95
300	0.86	0.72	0.58	0.32	0.52	0.88
400	0.69	0.46	0.52	0.13	0.41	0.81
500	0.26	0.22	0.14	0.07	0.20	0.54
600	0.21	0.10	0.11	0.05	0.15	0.41
700	0.15	0.08	0.09	0.03	0.10	0.10
800	0.09	0.05	0.06	0.02	0.06	0.07
900	0.04	0.03	0.03	0.01	0.03	0.03
1000	0.00	0.00	0.00	0.00	0.00	0.00
1100	0.00	0.00	0.00	0.00	0.00	0.00
1200	0.00	0.00	0.00	0.00	0.00	0.00

3.4. MATERIAL MODELS FOR REINFORCED CONCRETE

At present, there is no material model for reinforced concrete that accurately represents the material's behaviour in all possible scenarios during numerical simulations. Hence, knowledge and experience are needed in using the available material models and determining the relevant material input data. For the kinds of problems within the scope of the present report, a segregated approach, representing the concrete and the reinforcing bars separately, is commonly used. For the bond of concrete and steel, a rigid, not breakable, connection is usually assumed. The following sections briefly introduce a damage plasticity model for concrete and the Johnson–Cook model for reinforcement, respectively. These are two models relatively popular among analysts working in these kinds of applications.

A discussion on material characterization, with associated failure and acceptance criteria, can be found in, for example, Ref. [8]. An example of equivalent material properties of reinforced concrete to be used in materially linear analyses is presented in Appendix II.

3.4.1. Concrete damaged plasticity model

One example of a concrete model is the 'concrete damaged plasticity' model, in which material degradation is taken into account in compression and in tension [49]. Damage is associated with cracking and crushing. In scalar damage theory, the stiffness degradation is isotropic. Under uniaxial tension, the stress–strain relationship is:

$$\sigma_t = (1 - d_t)E_0(\varepsilon_t - \bar{\varepsilon}_t^p) \tag{38}$$

where

d_t is the tensile damage parameter;
E_0 is the undamaged modulus of elasticity;
ε_t is the tensile strain;

and $\bar{\varepsilon}_t^p$ is the equivalent plastic strain in tension.

In compression, correspondingly:

$$\sigma_c = (1 - d_c)E_0(\varepsilon_c - \bar{\varepsilon}_c^p) \tag{39}$$

where

d_c is the compression damage parameter;
ε_c is the compression strain;

and $\bar{\varepsilon}_c^p$ is the equivalent plastic strain in compression.

Compressive stiffness is recovered upon crack closure as the load changes from tension to compression, but the tensile stiffness is not recovered when the load changes from compression to tension. In Fig. 22, $\Gamma_t = 0$ corresponds to 'no recovery' as the load changes from compression to tension and $\Gamma_t = 1$ corresponds to 'complete recovery' as the load changes from tension to compression. It should be noted that the recovery stiffness assumption affects the bending vibration behaviour of the damaged structure.

3.4.2. Reinforcement steel

Non-linear material behaviour of reinforcement steel can be modelled using an elastic-plastic material model with von Mises yield criteria and isotropic strain hardening. The thermo-visco-plastic behaviour of steel can be represented by a Johnson–Cook constitutive equation. In that case, the von Mises flow stress σ_{eq} is expressed as:

$$\sigma_{eq} = (A + B\varepsilon^n)\left[1 + C\,\ln\left(\frac{\dot{\varepsilon}}{\dot{\varepsilon}_0}\right)\right]\left[1 - \left(\frac{T - T_{room}}{T_m - T_{room}}\right)^m\right] \tag{40}$$

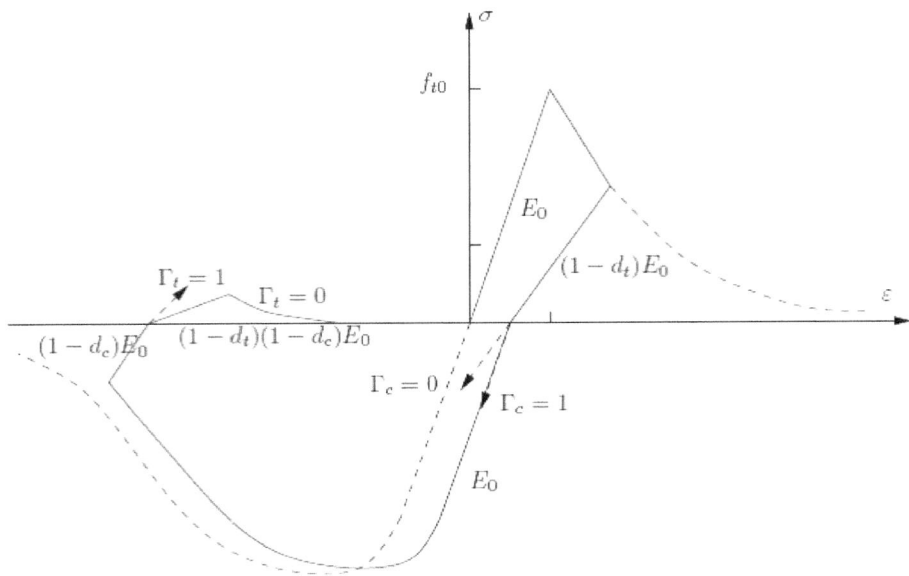

FIG. 22. Concrete damaged plasticity model.

where

ε	is the equivalent plastic strain;
$\dot{\varepsilon}$	is the strain rate (s^{-1});
$\dot{\varepsilon}_0$	is the reference plastic strain rate (s^{-1});
T	is the temperature of the material (°C);
T_m	is the melting temperature of the material (1399°C for steel);
T_room	is the room temperature (e.g. 20°C);
coefficient A	is the yield strength (MPa);
B	is the hardening modulus (MPa);
C	is the strain rate sensitivity coefficient;
n	is the hardening coefficient;

and m is the thermal softening coefficient.

4. STRUCTURAL RESPONSE ANALYSIS

4.1. GENERAL CONSIDERATIONS

Response analyses of SSCs range from non-linear detailed finite element analyses using explicit integration techniques to simpler analyses performed in database form using loading and capacity information. Decisions as to which methodologies to employ are based on a number of considerations, including the metrics of interest (maximum deformations versus time histories of response; consequences of failure, such as expected sizes of breaches of containment; etc.), the complexity of the geometry and plant layouts. The following discussion ranges from the least to the most sophisticated response analysis techniques. In the context of the evaluation, a reasonable approach is to attempt to screen out loading scenarios and their consequences in this same order of application. The implicit assumption is that as sophistication is augmented, the conservatism in the analysis is reduced.

As described in Section 2, impact consequences are traditionally categorized into local and global effects. Local effects are due to hard (or semi-hard) missile impacts, resulting in punching failure mode and they include penetration, perforation, scabbing and spalling. Global effects include overall axial, bending and shear effects in the structural elements between the impact area and support locations, global stability and vibration in the structure. First, the impact scenarios need to be identified. The impact scenarios are defined by missile type, mass, velocity, impact location and impact angle. Different impact locations may be defined for evaluating overall global and local integrity and vibration effects. In addition, different types of analysis tool may be needed for different types of impact analysis.

Second, the essential failure mode of the structure needs to be recognized and the model for analysis chosen accordingly. One practical way is to use simplified calculation methods. The selected analysis methods and models are to be verified to be able to simulate the essential failure mechanisms. The failure mechanism is dependent on the type of loading, geometry and dimensions of the structure, as well as on the location of the impact.

Material models used in the analyses need to be chosen to be able to simulate the essential material behaviour in a conservative and realistic manner. Material properties are dependent on temperature and strain rate. These phenomena may also considerably affect the collapse behaviour of the structure.

Comparisons of loading functions corresponding to aircraft crashes and an explosion are presented in Fig. 23. The short duration of a peak pressure transient due to a near field blast can clearly be seen in the right hand side plot. The analyst needs to be conscious that the time span of the different types of load being considered varies to a large extent, and to select the tools and models accordingly. Examples of strain rates for different types of loading are given in Table 8.

The accuracy of the calculation of the dynamic response induced in the interior of the structure (supporting locations of the systems and components) by a heavy impact load, such as that caused by an aircraft crash, depends mainly on the adequacy of the parameters used to describe dynamic material behaviour in the structure.

It is important to design building structures to exclude local penetration, as well as to transfer shock waves from the outer shielding structure to the interior. In the case of monolithic design and connections between the outer and inner structures, the high accelerations which may be expected on the outer shells in the vicinity of the impact locations are transferred directly to the inner structure. If the inner and outer structures are separated, the travelling shock waves may be attenuated significantly and filtered during the propagation all the way down to the foundation and from there to the supporting point of the equipment in the inner structure [50].

Considering the variety of shapes and structural concepts, it may be expected that there are always many (but not infinite) possible points of impact resulting in a maximum response in a certain area of the structure. The definition of the most representative impact locations is, however, more complicated in the case of box-shaped structures than in the case of axi-symmetric structures (Appendix II).

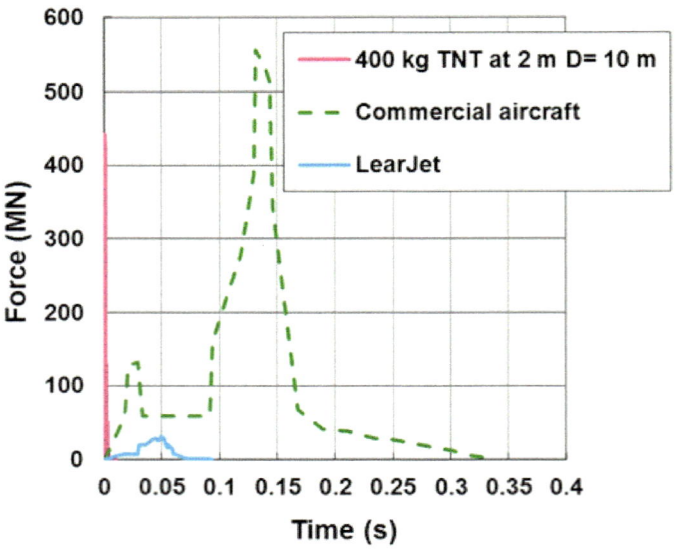

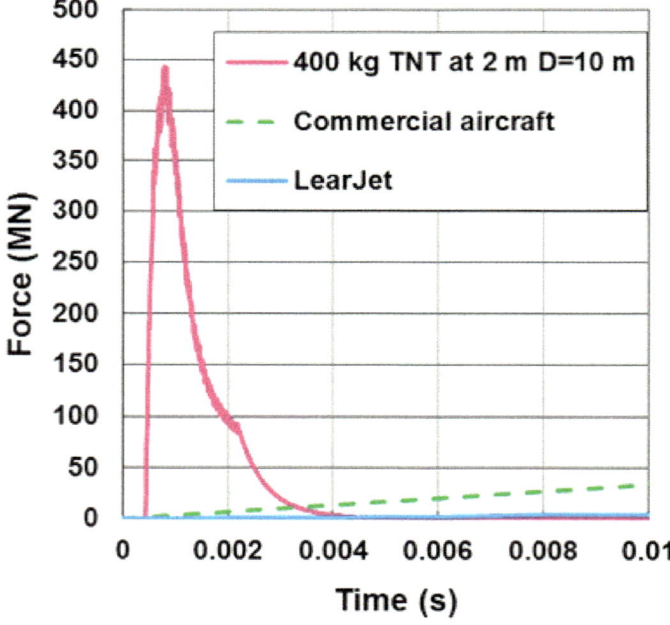

FIG. 23. *Comparison of loading functions due to aircraft impacts and a near field explosion (400 kg of trinitrotoluene at a distance of 2 m; diameter of loaded area: 10 m); a more detailed view is provided in the right hand side plot.*

TABLE 8. RANGES OF STRAIN RATE FOR VARIOUS LOADING TYPES

Loading type	Far field explosion	Soft impact aircraft impact	Hard impact	Near field explosion
Strain rate (s^{-1})	5×10^{-3}–1×10^{-1}	5×10^{-2}–2×10^{0}	10^{0}–5×10^{1}	10^{1}–10^{2}

In order to limit the extent of the calculations, it is recommended to specify a minimum number of regions and directions of impact on a building, which may cover all other impact locations within a practically acceptable range of scatter. The quickest way to specify the representative impact locations is a preselection of dominant locations by engineering judgement and demonstration of the correctness of the selection by case studies. Regarding global stability, the dominant impact locations will, however, correspond to the locations near the edges/corners in the upper regions of the corresponding building.

In aircraft crash analyses, sometimes a given loading cannot be applied to a structure entirely. For instance, the aircraft can hit the structure only partially. This aspect needs to be taken into account in the design process. Although local (engine impact) and global (fuselage impact) behaviour are considered separately in the design process for an aircraft crash, it should be kept in mind that they happen simultaneously or quasi-simultaneously. Even when perforation is precluded, the concentrated loading can produce significant scabbing, weaken a portion of the structure up to ten times the missile diameter in each direction and affect global behaviour [51, 52]. To avoid scabbing, the required concrete thickness or adequate transverse reinforcement needs to be provided. Both aspects need to be taken into account in the analysis. They are automatically taken into account in coupled analysis.

4.2. STRUCTURAL RESPONSE ANALYSIS PROCEDURES FOR IMPACT LOADS

4.2.1. General considerations

There are several possibilities to carry out dynamic analyses for impact loaded structures. The structural analyses can by divided into two categories: structural integrity analyses and vibration propagation analyses. The major analysis steps are as follows:

(a)   Selection of representative impact locations.
(b)   Load characterization (Section 2).

(c)   Calculation of model discretization (Section 4.2.3.1).
(d)   Boundary conditions (Section 4.2.4).
(e)   Application of loading (Appendices I and II).
(f)   Selection of material models (Section 3).
(g)   Structural response analysis (Section 4 and Appendix II):
    (i) Simplified methods are recommended for preliminary studies, comparative analyses and sensitivity studies (Section 4.2.2);
    (ii) Detailed analyses (Section 4.2.3).
(h)   Sensitivity study for essential calculation parameters (Section 4.2.3).
(i)   Assessment of performance (Section 5).
(j)   Induced vibrations and in-structure response spectra (Section 4.2.5 and Appendix II).
(k)   Detailed global and local analysis (considering in detail their global and local stability, as well as the real non-linear material behaviour of the loaded target) in order to ensure the penetration resistance and derivation of the induced vibrations (Sections 5.3.2.3 and 5.3.3).
(l)   Capacity check for the technological systems installed in the structures to demonstrate their sufficient design functionality under the induced loads (Section 5.3.2.3).

4.2.1.1. Historical perspective

In the past, the most effective and commonly used analysis procedures for the derivation of dynamic response results due to mechanical impact loading (aircraft crash, explosion, drop of heavy masses, etc.) was the time history method. The input excitation is a loading function (i.e. force time history) obtained by the assumption of an impact on a rigid object or by use of a verified method (considering the non-linear effects of the impacted target).

Since non-linear dynamic analyses could not be used in the past, due to insufficient computational capacity, by the end of the 1970s the VLF method was developed and qualified step by step for practical use [53].

A loading function derived under the assumption of an impact on a rigid body (rigid load functions (RLF)), when used in a linear-elastic analysis, provides very conservative dynamic response results. In order to consider the non-linear processes which have to be expected in the loaded area of the impacted object, a procedure for derivation of VLFs was established. Derivation of VLFs (acting in reality) on the basis of internal forces obtained for representative locations at the impacted object by non-linear calculations is presented in Appendix II.

4.2.1.2. Current status

Highly non-linear processes are expected in the impacted region of the structure, associated with significant local deformations. A rapid increase in stress and utilization of the capacity of the concrete and the reinforcing steel over the stress limits are expected as well. This process consequently also results in high energy dissipation and reduction of the forces transferred to the part of the global structure which still remains in a linear-elastic state. The global forces acting on the border of the partially non-linear zones of the structure are, therefore, accordingly reduced when compared with the loads originally applied by the missile.

The loading can be modelled by using either a coupled missile target computational model or a loading function calculated, for instance, with the Riera approach (Section 2.2.2). Some examples are given in Sections 4.2.3.3 and 4.2.3.4. In this type of modelling technique, depending on the size of the damaged area, it may only be necessary to model part of the structure as non-linear material. In such a case, the approach is similar to the VLF method.

The shape and duration of the loading function (the rise and decay parts, in particular) have a major influence on the character and frequency content of the dynamic response (response spectra) calculated with a global structural model. When evaluating the loading function and deriving the dynamic load factor (DLF) spectrum (Fig. 24), the frequency range considered in the linear–elastic dynamic analyses is to be taken into account.

The DLF is defined, for a linear system, as the ratio between the maximum deflection caused by a dynamic load and the maximum deflection caused by the same load applied statically (i.e. very slowly). For a single degree of freedom (SDOF) system:

$$m\ddot{x} + c\dot{x} + kx = f(t) = f \propto (t) \tag{41}$$

$$\text{DLF} = \frac{x_{max}}{x_{stat}} \tag{42}$$

where

$$x_{stat} = \frac{f}{k} = \frac{1}{\omega} \times \frac{f}{m}$$

Given a load time history *f(t)*, DLFs can be computed for a series of oscillators, each one characterized by a circular natural frequency ω and a damping ratio. The results can be represented as a DLF spectrum.

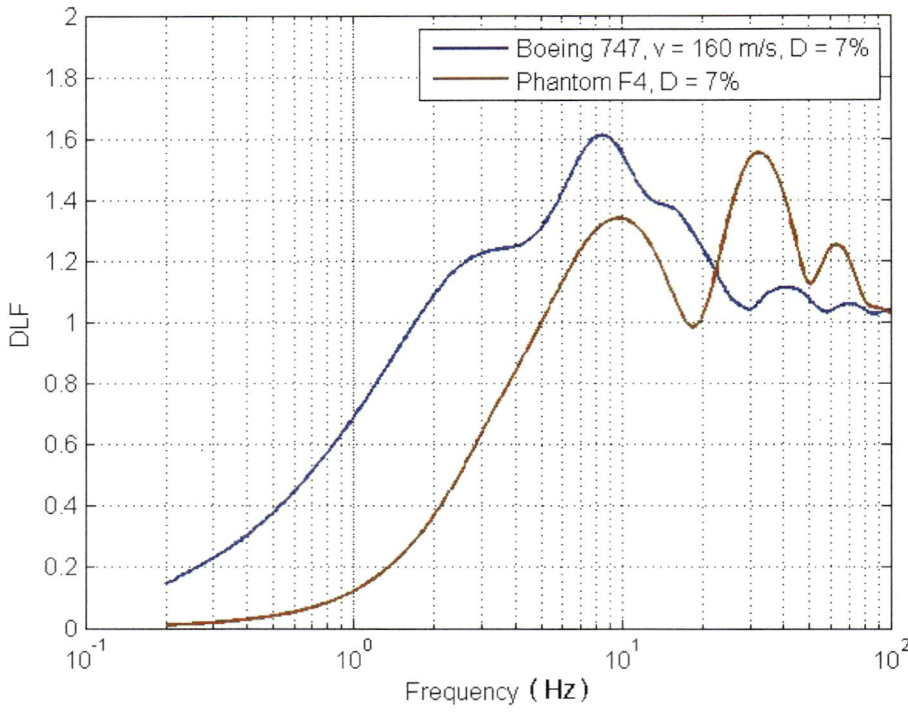

FIG. 24. Dynamic loading factors (DLFs) for representative aircraft.

In order to correctly consider the dynamic characteristics of an impacted structure, it is necessary, in the dynamic analysis, to take into account the contribution of all structural modes in the range of the significant frequencies that appear in the DLF spectrum of the load (Fig. 24). For military aircraft, this is approximately from 25 to 80 Hz, whereas for commercial aircraft, the range is approximately from 10 to 25 Hz. At high frequencies, the military aircraft case dominates, and at low frequencies, the commercial aircraft case dominates.

4.2.2. Simplified methods

In order to study both bending and shear failure of a reinforced concrete plate or shell impacted by a missile, at least a two degree of freedom (TDOF) model is needed, such as the Comité euro-international du béton model of Ref. [54]. This method, as described below, has been developed for reinforced concrete structures. Another approach to using simplified methods is reported in Ref. [55].

In Fig. 25, spring 1 and mass 1 are related to the global bending deformation of the plate, while spring 2 and mass 2 are used in modelling the local shear behaviour in the neighbourhood of the missile impact area.

The behaviour of element 1 (bending spring) is shown in Fig. 25 and the local behaviour related to the possible formation of a shear cone (shear spring) is shown in Fig. 26. The internal force in spring 2 is composed of the contributions due to concrete r_c, stirrups r_s and bending reinforcement r_b. Concrete behaves elastically until the displacement difference $u_{21} = u_2 - u_1$ reaches the value u_{cu}. Stirrups are assumed to break when the difference is $u_{21} = u_{su}$. The ultimate displacement connected to concrete deformation u_{cu} is very small but usually a large displacement difference is needed to activate a significant bending reinforcement contribution to the shear spring force. The bending reinforcement breaks when $u_{21} = u_{bu}$.

4.2.2.1. Stiffness, strength and effective mass of bending mode

In a cracked state, when concrete (in compression) and reinforcing steel still behave elastically, a bending rigidity coefficient can be determined by assuming a triangular concrete stress distribution over a top compressed zone with a depth of x [56]. If d is the effective slab depth (from the top) and the distance from the

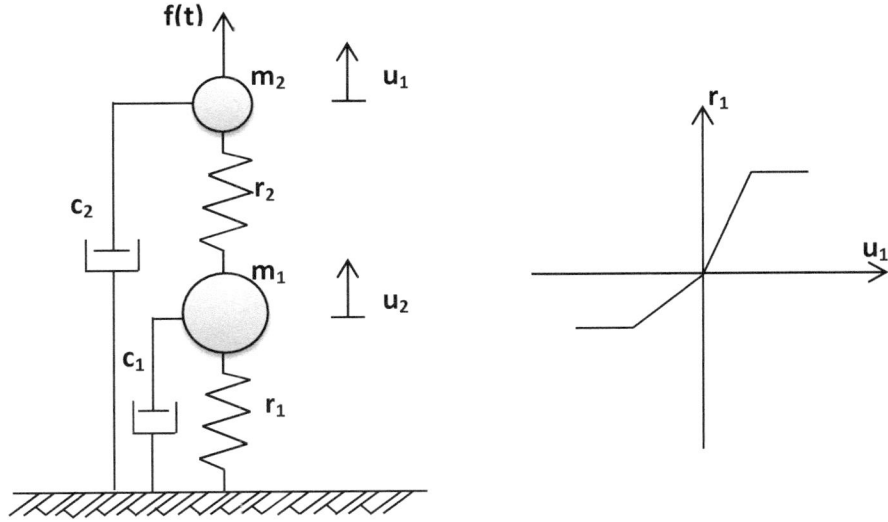

FIG. 25. A two degree of freedom impact model.

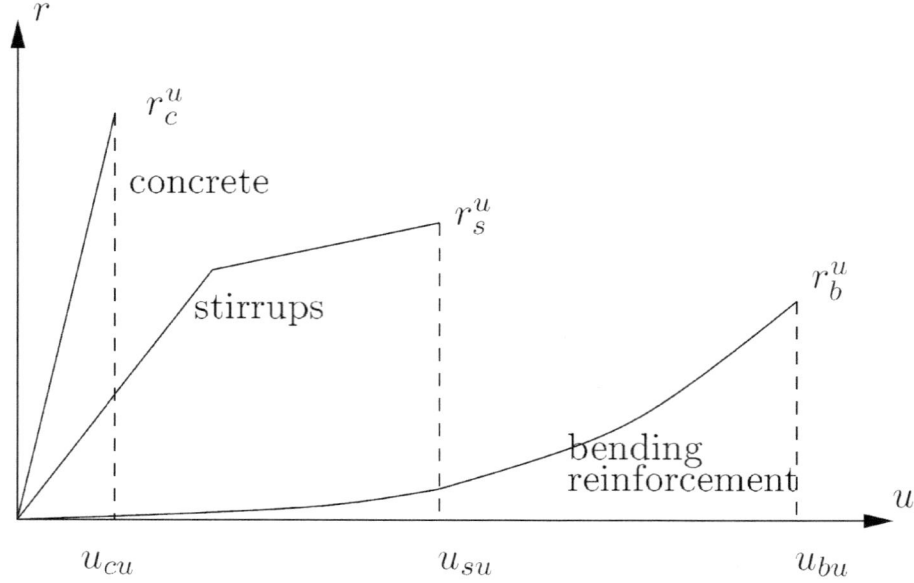

FIG. 26. *Local slab shear strength model showing the contributions of concrete, stirrups and bending reinforcement [56].*

neutral axis to the bending reinforcement is $d - x$, then the bending stiffness per unit width of cross-section is:

$$D = \left(d - \frac{x}{3}\right)(d - x)A_s E_c \qquad (43)$$

where

A_s is the reinforcement area;

and E_s is the Young's modulus of steel.

Introducing the ratio $n = E_s/E_c$, where E_c is the Young's modulus of concrete, the horizontal equilibrium equation in the absence of an axial load yields:

$$x = -nA_s \pm \sqrt{n^2 A_s^2 + 2nA_s d} \qquad (44)$$

The reinforcement ratio is defined as:

$$\rho_s = \frac{A_s}{d} \tag{45}$$

If the top 't' and bottom 'b' bending reinforcements are different, then the above values for x and D are determined for the loading direction x_b and D_b, and for the opposite direction x_t and D_t.

Additionally, the limit load and the effective mass are needed for the equations of motion of the TDOF system.

For a simply supported one way slab with a width of B and a span of L, the bending spring stiffness becomes:

$$k_b = \frac{48D \cdot B}{L^3} \tag{46}$$

The limit load obtained with a central yield line for a one way supported slab is:

$$R_p = \frac{4m_p B}{L} \tag{47}$$

Denoting as a the depth of the compressed zone in the slab cross-section, the equilibrium equations:

$$0.85a \cdot f_c = A_s f_y \quad \text{and} \quad m_p = A_s f_y \left(d - \frac{a}{2}\right)$$

yield a plastic bending moment per unit width of:

$$m_p = \rho_s d^2 f_y \left(1 - \frac{\rho_s f_y}{1.7 f_c}\right) \tag{48}$$

where

f_y is the yield stress of steel;
f_c is the compression strength of concrete;

and ρ_s is the reinforcement ratio.

The effective mass calculated with a piecewise linear deflection profile becomes:

$$m_e = \frac{1}{3} r \cdot h \cdot l \cdot b \tag{49}$$

4.2.2.2. Local behaviour

The local resistance of the slab to impact load is due to concrete, stirrups and bending reinforcement. The resistive force of concrete alone can be determined by assuming a shear cone with an angle of inclination α measured from the horizontal plane (Fig. 27).

The shear capacity is, thus:

$$F_s = F_{sc} + F_{sb} + F_{ss} \tag{50}$$

In the shear capacity formula:

$$F_{sc} = \tau_c h_p \pi \left(2r + \frac{h_p}{\tan \alpha} \right) \tag{51}$$

where α is the angle shown in Fig. 27.

The contribution of bending reinforcement is:

$$F_{sb} = \pi \left(2r + \frac{h_p}{\tan \alpha} \right) A_s f_y \sin \alpha \tag{52}$$

where

f_y is the yield strength of reinforcement;
A_s is the area of reinforcement per unit width (m²/m);

and the force due to bending reinforcement is assumed to act perpendicular to the shear cone surface.

Similarly, the effect of shear reinforcement can be evaluated by:

$$F_{ss} = \pi \frac{h_p}{\tan \alpha} \left(2r + \frac{h_p}{\tan \alpha} \right) A_{ss} f_{ys} \tag{53}$$

where A_{ss} is the area of shear reinforcement per unit area (m²/m²).

Perforation is initiated when the contact force equals the resisting force or $F_s = F$. The shear strength of the concrete cone $\tau_c = f_c / \sqrt{3}$ is based on Mohr's circle analysis in Ref. [57].

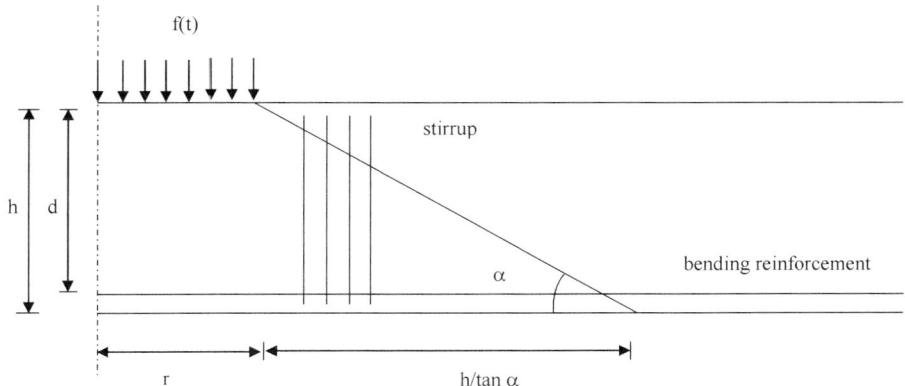

FIG. 27. Assumed punching shear cone.

The mass of the assumed punching cone is

$$m_2 = \tilde{\rho}\pi h \left[a^2 + \frac{a \cdot h}{\tan\alpha} + \frac{1}{3}\left(\frac{h}{\tan\alpha}\right)^2 \right] \qquad (54)$$

where

$\tilde{\rho}$ is the equivalent density of reinforced concrete;
α is the angle of the punching cone (measured from the horizontal plane);

and h is the slab thickness.

The mass number 1 in the TDOF model is

$$m_1 = m_e - m_2 \qquad (55)$$

An example of TDOF application is given in Appendix II.

4.2.3. Detailed analysis

Detailed analysis needs to be performed for the complex impactive and impulsive loading cases (e.g. aircraft crash analysis). The possible combinations of load definition, model setup, analysis type and material behaviour result in a series of four methods, which are shown in Table 9.

TABLE 9. COMPUTATIONAL APPROACHES

1	2	3	4
Riera load time function or verified load time function	Riera load time function or crash load time function	Riera load time function or crash load time function	Modelling of aircraft and target with contact interaction
Shell elements	Thin shell elements	Thick shell or solid elements	Solid elements
Modal	Implicit/explicit	Implicit/explicit	Explicit
Linear	Non-linear	Non-linear	Non-linear

The first method was developed in the 1980s. Nowadays, it is rarely used, as non-linear effects, which are relevant for the impact of a missile or an aircraft, are considered only indirectly. This method is still applicable if the structural behaviour is essentially elastic.

The second method uses thin shell elements to represent the structure. Thus, local damage effects due to penetration cannot be captured using this method. Its application is limited to load cases in which bending dominates the structural behaviour. If this method is used, a separate analysis needs to be performed for punching shear verification.

Methods 3 and 4 can capture both bending and punching behaviour. The methods are different with respect to the load definition and the simulation procedure. Method 3 represents a decoupled approach, where the missile is represented as a loading function $F(t)$. The loading function is defined assuming the missile crashes into an infinitely rigid target. The limitation of this method is that missile perforation cannot be modelled. The loading function has to be applied to elements (or nodes). Thus, the loading function cannot be applied to destroyed elements.

Method 4 integrates both models, the missile and the impacted structure, within one simulation. This enables the user to investigate and evaluate local and global structural effects simultaneously. The main advantage of the integral analysis is that it can capture missile perforation and follow the missile trajectory after the perforation, as well as possible damage to internal structures. The disadvantage of this comprehensive method is the large computational effort.

In summary, two approaches are used nowadays in finite element modelling:

(a)   The modelling of the entire system, including both missile and target with contact interaction between them during the impact (method 4);

(b)   A simplified approach using only a model of the target and the Riera loading function (methods 2 or 3).

The modelling of the entire system is more accurate. However, this modelling is much more time consuming and unpredictable. Additionally, the adequate missile model (e.g. aircraft) required for such modelling could be unavailable.

The second approach, using the Riera loading function, is widely considered as adequate in cases in which there is no or very limited damage of the target and a mainly perpendicular impact takes place [8]. However, this approach, coupled with an adequate material model, could also be used, at least as a rough estimation, for cases involving deep penetration and even perforation. As a criterion for assessing validity, a comparison of computed strains with ultimate strains can be used.

Extensive research conducted so far indicates that only explicit integration finite element analysis, with highly non-linear material constitutive models for concrete, provides an adequate analysis of severe structural damage with perforation [58]. The results of the same research programme show that, due to the complexity of the problem, successful simulation needs to be conducted by a highly experienced team in this specific area. The simulation needs to be performed following the steps below (taken from section 1.7 of Ref. [58]):

(a)   The first step of a detailed analysis is to analyse the problem and to define the most relevant results (local resistance to penetration/perforation or punching behaviour, global or semi-global flexural behaviour, combined flexural and punching behaviour, induced vibrations).

(b)   The second step is to choose an appropriate finite element code and the type of model (impacted structure and missile modelling), the type of elements and the material constitutive models. Lagrangian finite element analysis, using continuous beam, shell and solid finite elements, is most commonly used. However, discrete element methods and the smoothed particle hydrodynamics (SPH) method have gained popularity in recent research programmes.

(c)   The third step is to assess the main parameters (physical and numerical) related to the chosen results using sensitivity studies. The sensitivity studies need to be carried out, at a minimum, on the element size, the time step, the parameters of material constitutive models, the boundary conditions and the failure (or erosion) criteria. It should be noted that non-linear simulations have inherent uncertainties and sensitivity studies are to be used to define upper and lower-bound solutions.

(d) The fourth step is to calibrate the material constitutive model using representative test results. It is important to verify whether the simulation case is within the validity domain of the model.

(e) The fifth step is to compare the most important results (e.g. displacements, strains, response spectra, damage footprint and missile residual velocities) with the results obtained using alternative simplified methods. The simplified methods are often reduced to the analysis of a single parameter. Thus, there is a need to use a set of simplified models. This step is done independently of detailed analysis and can be performed at any phase. It is recommended to perform this step as soon as possible to have a rough idea about the expected results.

(f) The sixth step is to have an independent review of the analysis and obtained results. Taking into account the level of complexity and the range of uncertainties, an independent review is mandatory for this type of simulation.

4.2.3.1. Finite element model discretization

In the case of short duration impact loading, the type of model used for the idealization of the building structure and the discretization ratio of the model are fundamentally important for achieving realistic response results. It is not possible to define analytically the required optimal size of the finite elements to be used for an appropriate discretization of a structure. The finite element model needs to adequately represent the structural behaviour. A finer mesh is needed around suspected strain concentration areas. It is necessary to perform sensitivity studies to get the correct response. The model discretization requirements are different for the three main phenomena typical of impact analyses (flexure, punching, induced vibrations). It should be noted that, in general, all three phenomena occur concurrently during an impact. The following guidelines are elaborated from Ref. [58]:

(a) Simulation of the flexural structural behaviour due to an impact can be successfully achieved with shell elements. However, the element size influences the evaluation of the damage. Most commercial computer codes smear the damage over the element size. Hence, increasing the element size tends to smooth localized damage (such as cracking) and the computed maximums of local damage indices are reduced. In order to consider this effect, tensile cracking behaviour modelling needs to be adjusted according to the element size.

(b) Simulation of punching behaviour (modelling of penetration and perforation) requires solid three dimensional elements and mesh densities much higher than those used for simulation of flexural behaviour. Sufficient mesh density is to be defined using sensitivity studies. The extent of the area with fine mesh needs to be studied carefully and depends on the details of the reinforcement. The presence of transverse reinforcement reduces the size of the scabbed area and, consequently, the need for a large area with a fine mesh [1, 59].

(c) When simulating impact induced vibrations, it is to be considered that finite element models behave as low pass filters in relation to high frequency travelling waves. This filtering effect can, however, be demonstrated by means of parametric studies. In order to demonstrate the applicability limits of the finite element models for simulation of impact transfer of the induced waves in typical structures, extended parametric studies are given in Refs [60, 61]. Based on this study, performed for a wide range of discretization ratios on typical partial models (beams, walls and roofs), as well as for simple types of three dimensional building structure, the low pass filter sizes of the corresponding finite element types (beams, plates and solid elements) were specified, and empirical formulas for the definition of the allowable maximum element size depending on the frequency of the highest mode to be considered were defined.

In order to achieve sufficient accuracy of dynamic analysis results, sensitivity studies need to be performed. The following element sizes could be used for the first iteration [60]:

Beams, columns: $\dfrac{c_L}{12 f_n}$

Walls: $\dfrac{c_L}{22 f_n}$

Plates, floors, roofs: $\dfrac{c_{L,S}}{16 f_n}$

Box-shaped structures: $\dfrac{c_{L,S}}{16 f_n}$

Axi-symmetric structures: $\dfrac{c_{L,S}}{8 f_n}$

where

c_L is the longitudinal propagation velocity of concrete material;

c_S is the shear wave propagation velocity of concrete material;

and f_n is the upper limit of the frequency range of the loading function that needs to be considered.

The notation $c_{L,S}$ means that both longitudinal and shear wave propagation need to be considered when defining the element size.

$$c_L = \sqrt{E \frac{1-\nu}{\rho(1+\nu)(1-2\nu)}} \qquad (56)$$

$$c_S = \sqrt{E \frac{1-\nu}{2\rho(1+\nu)}} \qquad (57)$$

where

E is the Young's modulus (Pa);
ρ is the density (kg/m^3);

and ν is Poisson's ratio.

The smaller element size obtained by the above empirical formulas is to be used. Special attention should be paid to avoid the mesh sensitivity of the analysis results when carrying out finite element analyses with non-linear material models for reinforced concrete.

Regarding the modelling of impact induced vibrations using coupled analyses, the computed time history response of the structure needs to be passed through a low pass filter to preserve the essential characteristics of the structural response while eliminating numerical noise.

4.2.3.2. Damping

Traditionally, damping used in design or analysis of nuclear power plants is an experimentally determined factor which is used to make the results of linear-elastic analysis of dynamic systems match reasonably well with observed experimental results. Damping values are generally available for seismic analysis. For concrete structures, damping ratios in the range of 0.05 to 0.10 have been obtained. Damping trend curves are presented in Ref. [62]. However, damping ratios for seismic analysis are probably not valid for aircraft crashes.

Damping values for reinforced and pre-stressed concrete structures of nuclear facilities are given in Ref. [63] as a percentage of the critical damping, as shown in Table 10.

Damping values for linear analyses with hysteretic material behaviour are presented in the new German KTA 2201.3 [64]. Damping values given in ASCE 4-98 [62] are identical to the German ones, but additional information is added connecting the damping value to the ultimate strength of the concrete and yield of steel. The damping values from Ref. [62] are given in Table 11.

In practice, damping is to be introduced into the calculation within the limits of the selected analysis method:

(a)   Viscous damping. Viscous damping forces are proportional to velocity and, thus, inadequate for describing damping in many real structures since damping is, in many cases, independent or weakly dependent on frequency.
(b)   Complex stiffness damping. In order to avoid the frequency dependence of damping, the uncoupled normal mode equations can be solved in the frequency domain using complex stiffness damping rather than viscous damping.
(c)   Modal damping. Modal damping is applied in mode superposition analyses when linear models are used. The advantage of this methodology is that different damping values can be applied for different eigenmodes.

TABLE 10. DAMPING RATIOS FOR CONCRETE STRUCTURES, PERCENTAGE OF CRITICAL DAMPING

| Class | Construction | |
	Reinforced concrete (%)	Pre-stressed concrete (%)
A1[a]	2	1
A2[b]	4	2
A3[c]	7	5

[a]   Class A1: normal load cases, permanent and moving loads.
[b]   Class A2: exceptional load combinations.
[c]   Class A3: combined exceptional load case probability less than 10^{-4}/a.

81

TABLE 11. DAMPING RATIOS FOR LINEAR ANALYSES, PERCENTAGE
OF CRITICAL DAMPING

Structure type	Normal operation (%)	Design basis event (%)	Design extension event – level 1 (%)
Reinforced concrete	4	7	10
Pre-stressed concrete	2	5	7
Welded and friction bolted steel structures	2	4	7
Bearing bolted steel structures	4	7	10
Welded aluminium structures	2	4	—[a]

[a] —: No data available.

(d) Rayleigh damping. A common damping model in non-linear finite element codes is Rayleigh damping. Damping is assumed to be proportional to a combination of the mass and the stiffness matrices:

$$C = \alpha M + \beta K \tag{58}$$

where the two Rayleigh damping factors α and β can be evaluated if the damping ratios ξ_m and ξ_n associated with two specific frequencies (modes) ω_m and ω_n are known. In practice, the damping factors are calculated by choosing a damping ratio ($\xi_m = \xi_n$) at selected frequencies (ω_m and ω_n). It should be noted that the damping ratio in the frequency range between ω_m and ω_n is lower than ξ_m. More important from the conservative point of view is that outside this range the damping ratio increases considerably and the analysis results are over damped.

4.2.3.3. Non-linear model loaded with a force–time function

One practical way of analysing a rather complicated structure is to apply time dependent loading functions, at least partly, to a non-linear analysis model. This methodology can be applied both for pressure transients due to explosions and aircraft impacts. Impact load due to an aircraft crash can be modelled using a loading function defined by the Riera method. As described earlier, the non-linear part of the model needs to be large enough. In the case of a double wall structure, all of the outer walls could be modelled using non-linear material models.

In principle, both the structural integrity of the impacted outer wall as well as the induced vibrations can be assessed using the same calculation model. In many cases, a shell element model is practical and, in that case, an additional punching capacity assessment is needed. To assess punching behaviour using finite element analysis, solid three dimensional elements are needed.

4.2.3.4. Coupled missile–structure model

Aircraft crash response analyses can be effectively performed using coupled analysis models, including both the impacted structure and the crashing aircraft. The time histories of the crash behaviour of the aircraft fuselage and the engines, and the time history of the reaction forces in the impacted regions, as well as the behaviour of the module building can be determined in one calculation run.

Simulating a crash using an aircraft model impacting the concrete structural response likely results in higher frequency content in the building response than is shown by using an equivalent load function obtained using the Riera force time history method [7]. More comprehensive (and realistic) treatment of the non-linear processes (e.g. extending the non-linear concrete modelling further from the impact area, such that all the concrete cracking is captured in response modelling) tends to reduce the response amplitudes in the high frequency range. It is known that the results obtained by means of the equivalent load function method may be quite sensitive to the assumptions associated with loading area and timing of load application (e.g. consideration of wing sweep angle). This needs to be considered if proposed simplified linear approaches are used, such that the high frequency characteristics are somewhat captured.

This type of coupled model analysis may be suggested only if the design data related to the aircraft are sufficiently clarified. The missile target interaction analysis method is described in Ref. [8]. The coupled aircraft–structure model analysis could be undertaken for benchmarking the results obtained using the equivalent loading function procedure. In some cases, the non-linear structural models can increase the high frequency content [65].

4.2.3.5. Finite element modelling of aircraft impact

Owing to the high non-linearity (both material and geometrical) of the impact process, explicit integration of finite element codes is usually used [58]. The most critical point is the selection of an appropriate material model that could adequately represent the behaviour of the concrete target, including effects such as spalling, scabbing, tensile cracking and compression crushing. The penetration of the target and/or its perforation also needs to be modelled. In this respect, the best results could be achieved when specialized user-defined material models and/or damage criteria are introduced into these commercial codes. These models are usually based on the extensive experimental and/or theoretical work in this area.

The simulation of penetration and/or perforation of the target could be achieved using finite element erosion (deletion) based on some predefined criteria, such as strain/stress limits or assessed damage. This erosion capability could be included in material model formulation or as an additional feature. However, the introduction of finite element erosion also produces some unwanted effects:

— The results depend on the selected erosion criteria, which are not directly based on the real (physical) material properties.
— Erosion leads to instantaneous change in the balance of forces in the surrounding finite elements, which could produce unwanted oscillations, shock waves, etc.
— Erosion also leads to changes in mass properties and to non-conservation of energy during simulations.

The SPH method has, thus, gained popularity in recent research due to its ability to model perforation of the target without including material erosion. However, this method has a limited capacity to model concrete reinforcement. Another setback of this method is the very high computational time, usually several times more than for 'classic' finite element analysis.

It should be noted that explicit integration finite element codes have one very important limitation: they are only conditionally stable. The integration time step needs to meet the Courant condition: the maximum allowed time step is tied to the minimum time needed for the stress waves to pass through the smaller finite element used in the mesh [66]. This limitation could lead to many hours, even days, of computer time for simulation of a few seconds of an impact in a large finite element model.

4.2.4. Boundary conditions and soil–structure interaction

Soil–structure interaction needs to be considered in the analysis of global behaviour of the structure. Soil–structure interaction may be modelled by simplified linear or non-linear approaches, which properly account for the effects of stiffness of the supporting soil and energy dissipation due to radiation damping and soil material damping, as well as for geometric non-linear effects at the structure/foundation interface with soil [59, 62]. Equivalent spring constant and damping values for a circular and for a rectangular base are given in Ref. [59].

Linear models of soil–structure interaction (i.e. using linear springs at the base of the structure) are appropriate when non-linear effects (uplift and sliding) are predicted to be minimal. The area with springs in tension is required to be less than 30% of the total base area in order to validate the linear model representing the soil–structure interface. Nevertheless, if the tensile area obtained with this linear approach is more than 30%, a non-linear time history analysis, an energy-based approach or other alternative justified approach can be used to demonstrate that 70% of the total interface area remains in contact. If one of these alternative methods is successful, the linear model is validated. If not, the model needs to be non-linear [67]. Equivalent spring constant and damping values for a circular and for a rectangular base are presented in Ref. [59].

4.2.5. In-structure acceleration and displacement response spectra

In-structure response spectra can be calculated as follows:

(a) Calculating the dynamic response of the building under considered loading cases.

A response spectrum represents the maximum acceleration, velocity and displacement of an SDOF oscillator as a function of frequency and damping. In general, equipment items having a small mass when compared to that of the supporting structure will not produce noticeable interaction effects on the response. These equipment items can usually be assessed by means of a separate (uncoupled) analysis, using the in-structure response spectra or excitation time histories computed for the location of the equipment supports in an analysis of the main structure. For these cases, the computational model of the main structure only needs to account for the actual mass distribution, including the supported equipment. However, massive equipment items, such as the reactor vessel, could have a dynamic interaction with the main structural system (dynamic coupling) and, consequently, modify the response with respect to the response computed as if they were just attached masses. The mass and stiffness of such large items must normally be modelled as part of the model of the main structure [68, 69]. The dynamic response of the structure, due to the impact loading, needs to be calculated using appropriate finite element models. Different types of modelling approach, as described in Section 4.2, can be utilized. The process needs to be repeated for each loading condition.

The VLFs (Appendix II) discussed above can be used as a basis for calculation of the dynamic response (response time histories of acceleration and displacements, and corresponding in-structure response spectra). It is expected that the acceleration response spectra obtained using the VLFs, instead of the load functions derived assuming a rigid target (RLF), will result in significantly reduced response results. The reduction ratio will be comparable to the results obtained for a pressurized water reactor type building structure in the example in Appendix II. The starting point for the investigations is to be based on a comparison of results from RLFs and VLFs using the same detailed finite element model.

To perform the dynamic response analyses, the same commercial finite element codes as in the field of structural analyses can generally be used.

(b)   Calculating the response spectra as post-processing of the dynamic response, at representative locations of the structural model where equipment and components of interest are located. Recommended damping values are 2%, 4%, 7% and 10% of critical damping.

The two horizontal and the vertical response spectra can be computed from the time history motions of the supporting structure on the various floors or equipment support locations of interest.

(c)   Defining the relevant frequency range of acceleration spectra, for assessing structural stability and integrity of equipment and components. This range is based on the displacement spectra.

It can be stated that the response spectra due to (military) aircraft crashes are, in comparison to further external event loading cases, characterized by much higher acceleration values, especially in the high frequency range. However, it can be observed that the associated spectral displacement values, beginning at a certain frequency range, reach very low values. Of crucial importance for defining the relevant frequency range of the spectra is, thus, the knowledge of corresponding displacement spectra. By evaluating the displacement spectrum together with the acceleration spectrum, it is possible to define the relevant frequency span f_R of the spectrum, considering the displacements which are allowable for the components from the engineering point of view (e.g. 1 mm). It should be noted that allowable displacements are those which can usually be absorbed by clearances or elasto-plastic deformations between the component and its support. Only the parts of the acceleration response spectra up to the relevant frequency f_R yield the representative acceleration level b_R, which may be used as a constant (quasi-static) acceleration value for aircraft crash design of components (Fig. 28 [70]).

(d) For functionality of equipment and components, the whole range of frequencies needs to be taken into account (Section 5.3.2.3).

(e) Smoothing of the response spectra, and broadening and clipping of the peaks.

Uncertainties in structural frequencies and material properties are taken into account by broadening the peaks associated with each of the structural frequencies. One method for broadening the peaks is presented in Ref. [68]. Methods for broadening and clipping are introduced in Refs [64, 71, 72].

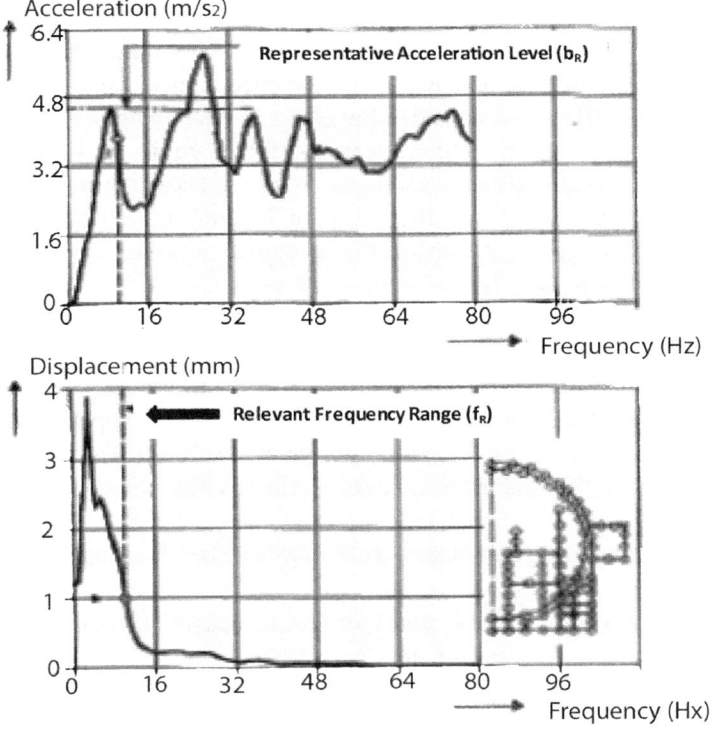

FIG. 28. Determination of the relevant frequency range and the representative acceleration level (example for 2% damping) (courtesy of IASMiRT [70]).

4.3. STRUCTURAL RESPONSE ANALYSIS FOR EXPLOSION LOADS

4.3.1. General considerations

The basic premise is that for the types of SSC of major interest, the fundamental periods of the supporting or sheltering structures are far longer than the time frame of the pressure loading conditions and that only the behaviour of local elements is subject to dynamic amplification in response to these loads; that is, the time frame of response for the overall structure is well beyond the time frame of the excitation. Hence, methods of analysis have been developed focusing on the evaluation of simplified representations of structural elements, such as equivalent single (or low number) degree of freedom models.

A most important consideration, for which considerable efforts have been made, is the non-linear energy absorption capacity of structures and structural elements. This is an essential aspect of the evaluation, as well as of the design, of blast loading resistant structures. The load lasts for a very limited time and failure takes place when deformation capacity is exceeded. Consequently, publications, such as Refs [9, 10], spend considerable effort in describing simplified models of ductile structures and structural elements, for example, reinforced concrete walls and slabs, structural steel beams and columns. These approaches generally simplify the representation of the pressure loading time histories, creating impulse equivalent representations of the combined pressure wave effects, to be applied to the local models of the elements of interest.

Several approaches have been developed and applied extensively to the analysis and design of buildings subjected to blast loading [9, 10]. Generally, these approaches progress from an evaluation of local effects (e.g. performance of walls and roof diaphragms), to the evaluation of supporting structures (e.g. beams, columns and foundations), to the evaluation of the overall stability of the structure itself. Load paths are to be evaluated in this sequential process.

4.3.2. Material properties of structures under blast loading

Material properties need to consider the specificity of blast loading and the requirements of simplified design approaches:

— Nominal strength increase factors may be used to reflect the actual strength of the structural elements. For concrete properties, these factors provide properties over the 28 d minimum values. For structural steel, the factors account for the margin over the minimum specified values. In the absence of actual site measurements, an increase of 10% can be used, if no degradation is observed [9, 10, 15].

— Dynamic increase factors may be used, which incorporate the effect of a larger material strength as a function of strain rate. Dynamic increase factors are dependent on the type of stress (flexure versus shear) and on the time it takes to develop the responses (see Section 3). Approximate values have been recommended in Ref. [9] based on perceived strain rates from explosive size and distance to the source. These are appropriate for simplified evaluations.

— Strain hardening. It is important to take into account the actual stress–strain material behaviour, either directly in the analysis of the equivalent stiffness models or through equivalent elastic-perfectly plastic models.

— Deformation limits. Failure is most often defined by deformation limits, which allows an evaluation to account for the significant effect of non-linear energy absorption and ductility.

Appendix III gives examples of calculations of structures with blast loads that illustrate these concepts.

4.3.3. Simplified method

4.3.3.1. Simplified versus complex models

In many cases, structural components subject to blast load can be modelled as an equivalent SDOF mass–spring system, with a non-linear spring (Fig. 29). In creating an equivalent SDOF structure, it is to be realized that the real structure is a multi-degree of freedom system in which every mass particle has its own equation of motion. Thus, to simplify the situation, it is necessary to make kinematical assumptions about the response and, in particular, on how to characterize global deformation in terms of a single point displacement.

Complex models usually involve non-linear finite element computations, in which non-linear geometric and material behaviour is considered and failure criteria implemented. Non-linear finite element computations may start from the load function definition, that is, predefined load functions that vary over space and time, and that are applied to the SSC of interest; or they may start from the explosion or blast itself, that is, from the position, amount, geometry and nature of the explosive material. In the latter case, the blast wave generation, propagation up to the structure and the interaction with it also need to be modelled. This normally requires the combination of Lagrangian and Eulerian processors in the computer code. In many commercial finite element codes, the pressure waves generated by the release of chemical energy in an explosion can be modelled by using, for example, the Jones–Wilkins–Lee equation of state. Special care needs to be paid to verify the calculation method in simulating pressure loads due to

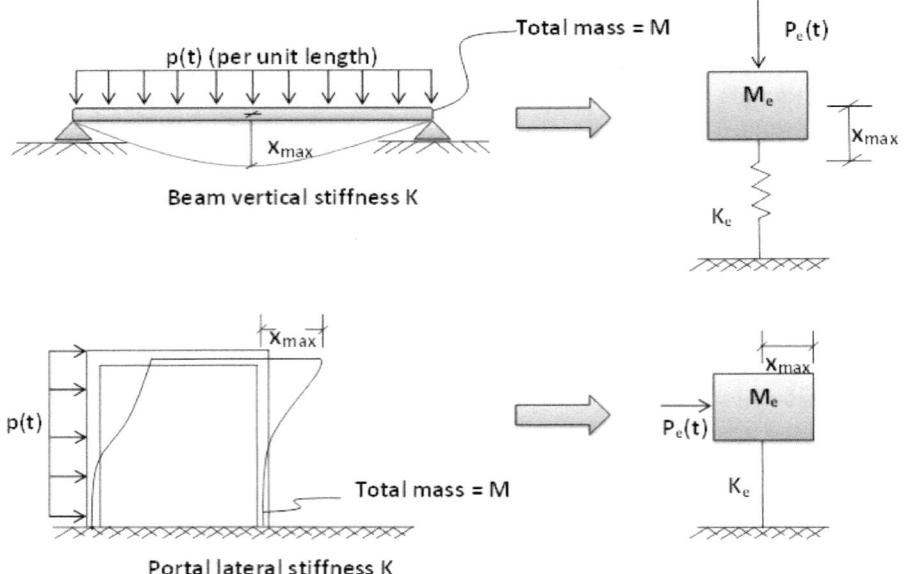

FIG. 29. Real and equivalent structural systems.

a near field detonation. Simulation of highly non-linear material behaviour is a challenging task. Strain rate dependency of material properties, as described in Section 3, needs to be taken into consideration. For extremely large strain rates (shock propagation), there are several non-linear material models available. Hydrodynamic material models using Mie–Grüneisen equations of state can be used. The SPH approach provides one possibility of simulating the breach through a structure. One numerical simulation study with experimental results is presented and discussed in Ref. [73].

Analyses with these sophisticated techniques are usually performed when simpler methods, generally with unquantifiable conservatisms, are likely to lead to conclusions of adverse failures. Other likely applications are for benchmarking simpler methods, especially when dealing with complex geometry or not easily modelled situations, such as explosions in confined or partially confined spaces.

Finite element methods for explosion loads are not further developed in this report; the focus is on simplified methods. Strictly speaking, simplified methods are not just calculating the response. Owing to the nature of the problem, elements of the capacity evaluation need to be introduced to permit their application.

4.3.3.2. Development of an equivalent single degree of freedom system

The mass and dynamic load of the equivalent SDOF system (Fig. 30) are based on the component mass and blast load, respectively; the spring stiffness and yield load are based on the component flexural stiffness and load capacity. The properties of the equivalent SDOF system are also based on load and mass transformation factors. These factors are calculated such that the SDOF system and the represented component have equal kinetic, work and strain energies at each time, assuming that the SDOF system deflection is always equal to the component maximum deflection and that the component maintains an assumed deformed shape as it responds to a blast load.

The basic equation of motion for an SDOF system under blast load is as follows:

$$M_e\ddot{x} + R_e(x) = F_e(t) \tag{59}$$

where

M_e is the effective mass;
R_e is the stiffness dependent effective resistance function;

and $F_e(t)$ is the effective blast load history.

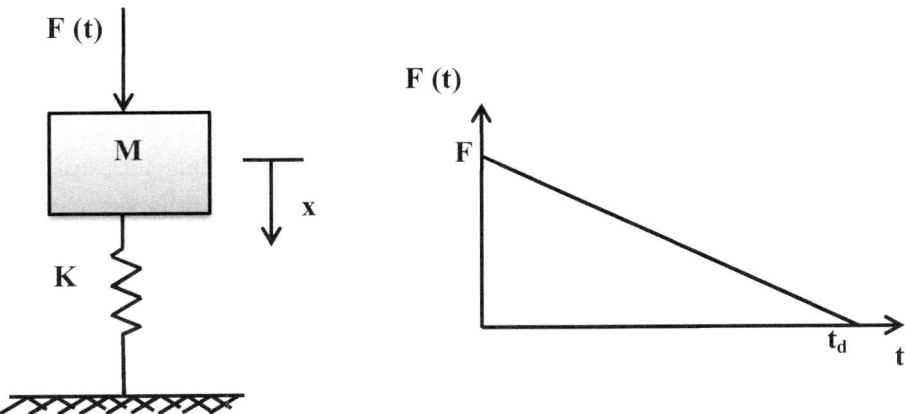

FIG. 30. A single degree of freedom elastic structure subject to an idealized blast pulse.

Comparing with the basic equation motion for an SDOF system under general dynamic loading, it should be noted that for blast loading the damping term can be ignored.

Under blast loading, the equivalent SDOF system deflects beyond the yield deflections and the formation of plastic hinges changes the deflected shape. This is taken into account using transformation factors and Eq. (59) becomes:

$$K_M M_c \ddot{x} + K_R R_c(x) = K_L F_c(t) \tag{60}$$

where

K_M is the mass transformation factor;
K_R is the resistance transformation factor;
K_L is the load transformation factor;
M_c is the mass of the blast loaded element;

and R_c is the resistance of the blast loaded element.

It can be shown that the resistance factor K_R has to always equal the load factor K_L [7]. This equation can then be simplified if the load and mass transformation factors are combined into a single load mass factor K_{LM}:

$$K_{LM} M_c \ddot{x} + R_c(x) = F_c(t) \tag{61}$$

where $K_{LM} = K_M/K_L$.

This equation is solved to determine the deflection history of the equivalent SDOF system, which is equal to the maximum defection history of the blast loaded element.

Basically, there are three different regimes regarding the structural behaviour under blast loading, which correspond to three different types of solution for Eq. (61):

(a)   The positive phase duration is long compared to the natural period of vibration of the structure. In this case, the load may be considered as being constant while the structure attains its maximum deflection. This can be a case derived from a very strong blast source at a great distance. Such loading is referred to as quasi-static or 'pressure' loading.

(b)   The positive phase duration is short compared to the natural period of the structure. In this case, the load has finished acting before the structure has had time to respond significantly. The maximum deformation of the structure occurs well after time t_d in Fig. 30. In this case, the structure is subjected to 'impulsive' loading (i.e. a momentum is transferred to the system, which translates into an initial velocity of the mass M in Fig. 30).

(c) The positive phase duration is similar to the natural period of the structure. In this case, the assessment of the response is more complex, possibly requiring a complete solution of the equation of motion (as mentioned in Eq. (59), the damping term is neglected), though it is often possible to have a reasonable approximation to the response by using results obtained for impulsive or quasi-static loading. In this case, the structure is said to be subjected to 'dynamic' loading.

The entire design process using SDOF systems is shown in the flow chart in Fig. 31. Examples of calculations illustrating the design of structures for blast loads are presented in Appendix III.

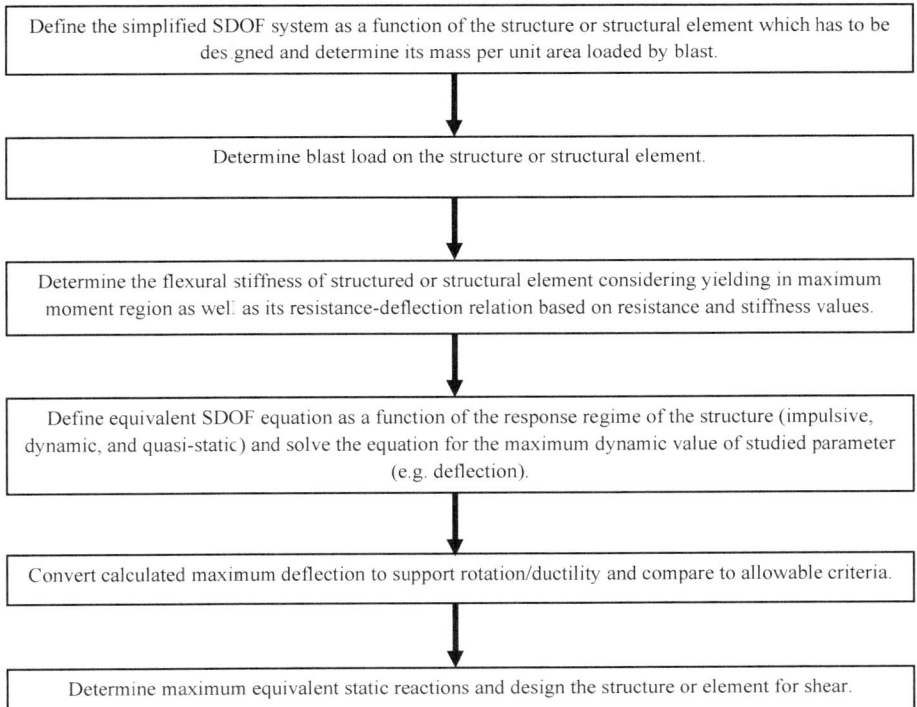

FIG. 31. Design process using single degree of freedom systems.

4.3.3.3. Pressure–impulse diagrams

Pressure–impulse diagrams are used in blast resistant design for quick assessment of structural elements or structural systems. This approach takes advantage of the non-dimensional scaling characteristics of the phenomenon to define the onset of structural damage as a function of the pulse loading parameters (Fig. 32).

The information necessary to construct a pressure–impulse diagram includes the equivalent static and dynamic characteristics of the structure being evaluated (i.e. those needed to build an SDOF equivalent system) and the shape of the loading pulse (e.g. triangular, rectangular). Then, for each pair of peak pressure P_{max} and impulse of the pressure distribution I, the maximum deflection is computed using an SDOF system. The computed maximum deflection is used to assess the expected level of damage.

It should be noted that once the peak pressure and pulse shape are fixed, pulse duration is proportional to the given impulse. Hence, for large impulses, the response will enter into the quasi-static regime, where the maximum dynamic deflection is equal to a constant factor times the static deflection (the maximum amplitude of load P divided by the equivalent stiffness K). For elastic systems, this factor is equal to two. For inelastic systems, the factor is less than two. As discussed earlier, this regime corresponds to loading functions where the time

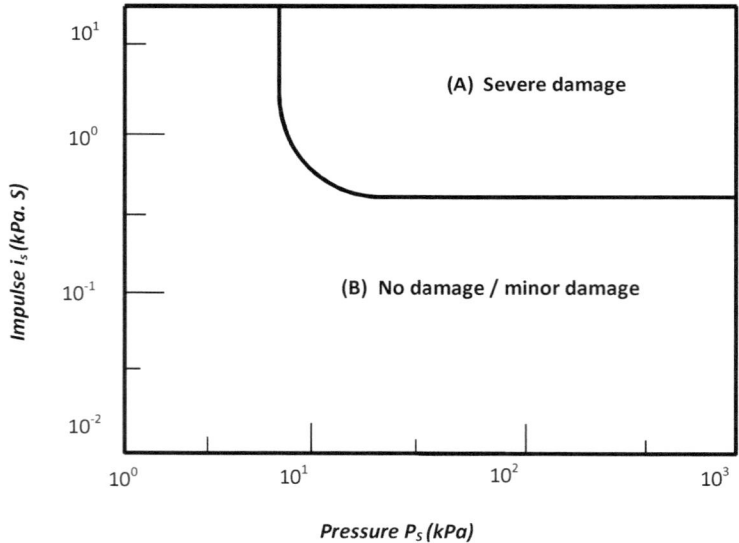

FIG. 32. A typical pressure–impulse diagram.

of application is long compared to the fundamental period of the structure. Once within this regime, the maximum deflection is dependent only on the stiffness characteristics of the system and the amplitude of the loading function (i.e. independent of the mass of the system and the total duration of the pulse).

On the other hand, once the impulse and the pulse shape are fixed, pulse duration is inversely proportional to peak pressure. Hence, for large peak pressures, the response will enter into the impulsive regime. As discussed earlier, this regime corresponds to loading functions where the time of application is short compared to the fundamental period of the structure. Once within this regime, the maximum deflection is dependent only on the impulse value and the dynamic characteristics of the system.

Consequently, the part of the pressure–impulse diagram corresponding to these two first regimes of the response can be constructed relatively easily.

A third regime is intermediate to the other two. It is the dynamic loading regime. In this regime, numerical analysis needs to be employed to calculate the maximum response. Again, however, it is relatively straightforward to do so, once the loading function is known and the equivalent SDOF representation of the structure or structural element has been built.

Failure is most often defined by deformation limits, which allows the evaluation to account for the significant effect of non-linear energy absorption and ductility. For analysis using pressure–impulse diagrams, curves of equal deformation, in terms of ductility factors can be developed in order to facilitate the work of the engineer making the assessment.

4.4. STRUCTURAL RESPONSE ANALYSIS FOR THERMAL LOADS

4.4.1. Analysis principle

The time–temperature curves of the standardized fire tests, such as ISO 834 [74] or ASTM E199 [75], have been created to resemble the severe exposure from a room fire. The classical way of assessing fire resistance consists of testing individual structural elements in a furnace, with burners producing a heating curve according to these standards. The time span to collapse of the structural element is determined and a 'rating' is assigned to the element accordingly. Using just the fire resistance rating based on such curves for a design meant to withstand external fires is questionable for several reasons:

— The standardized curves are theoretical, with no link to real fires;
— The curves do not consider actual fire loads, the ignition phase, ventilation conditions, etc.;

— The temperature is considered to be uniform, even for large spaces, and it never goes down.

Another problem with using the standard test results is that they provide little information to support the structural analysis of real fire scenarios. Having a fire barrier component rated for 90 min, for instance, does not mean that the component could perform its function for at least 90 min of a real fire. Nor does it mean that it would fail at that time. As a result, an analysis of the structural response against thermal load requires computational assessment of the thermal and mechanical behaviours.

Two different approaches can be used for the failure criteria. The first approach is to calculate the temperature distribution inside the structure as a function of time and to assume a simple failure criterion based on temperature. Some criteria of this kind are presented in Section 5.5.

The second, more demanding approach is to calculate the temperature distribution and the resulting mechanical response, taking into account the degradation of the mechanical properties, strains induced by thermal expansion and the resulting reduction of load bearing capacity. When investigating the thermal effects, the mechanical effects due to the initial event also need to be taken into account.

The computation of temperature distribution inside the fire barrier components and structures can be performed using either the semi-analytical methods for one dimensional heat conduction or numerical solutions of the heat conduction in one, two or three dimensions. Application of the semi-analytical methods for the SSCs is explained in Section 5.5.2.2.

4.4.2. Boundary conditions for thermal analysis

The thermal analysis is used to calculate the evolution of temperature within a target of interest, such as a building structure or safety relevant component. In practice, it means solving the numerical form of the heat conduction equation within the structure or component of interest. The boundary condition of the thermal analysis is the net heat flux, which is the sum of convective and radiative heat fluxes:

$$\dot{q}''(t) = \dot{q}''_c(t) + \dot{q}''_r(t) = h_c\left(T_g - T_s\right) + \varepsilon\dot{q}''_{r,in} - \varepsilon\sigma T_s^4 \tag{62}$$

where

h_c is the convective heat transfer coefficient ($kW \cdot m^{-2} \cdot K^{-1}$);
T_g is the local gas temperature (K);

T_s is the local surface temperature (K);

σ is the Stefan-Boltzmann coefficient;

ε is the surface emissivity;

and $\dot{q}''_{r,in}$ is the incident radiative heat flux (kW/m²) yielded by the fireball/pool fire radiation correlations or the radiation model of the fire CFD code.

A conservative estimate for h_c would be 0.02 kW · m⁻² ·K⁻¹ but more accurate values can be obtained using the heat transfer literature correlations for natural and forced convection. In general, the convection on the heated side of the structure needs to be considered forced, and the cooling of the cold side is caused by natural convection.

Some finite element codes require that the boundary condition of the heat transfer calculations be specified in terms of ambient temperature. In these cases, a useful quantity of information exchange is the so-called adiabatic surface temperature, which is defined as:

$$\dot{q}''(t) = h_c\left(T_{AST} - T_s\right) + \varepsilon\sigma\left(T_{AST}^4 - T_s^4\right) \qquad (63)$$

It is important to make sure that the values of h and ε in the fire and thermal analyses are consistent. The adiabatic surface temperature is often considered a numerical equivalent of the plate thermometer measurement used in experimental studies of fire resistance. The adiabatic surface temperature would be measured as an output quantity of the fire analysis and used as an input quantity for the structural analysis in place of ambient temperature.

5. ASSESSMENT OF PERFORMANCE AND ACCEPTANCE CRITERIA

5.1. GENERAL CONSIDERATIONS

5.1.1. Structures

The structures have three safety functions in the protective design against impactive and impulsive loading:

(a)   Containment of radioactive material or leak tightness (release control);

(b)   Mechanical protection of safety relevant SSCs inside the protective structures, including limitation of vibration;

(c)   Structural integrity (overall stability, no collapse, no perforation).

These functions can be fulfilled by two different structures (the shielding structure and the containment) or can be lumped into a single protective/ containment structure.

'Structural performance' refers to the degree of fulfilment of the safety functions. The major steps for structural performance assessment are as follows:

(a)   Identifying the global and local failure modes at different locations;

(b)   Checking the global stability;

(c)   Checking the structural integrity of the outer wall against appropriate criteria for local and global behaviour;

(d)   Assessing the in-structure response spectra at critical locations in order to ensure the functionality of the equipment considered.

The most relevant parameter to define structural performance is structural damage. In line with Ref. [76], three levels of structural performance, with associated damage levels, are defined in this report:

(a)   DBE: no damage (quasi-elastic behaviour).

(b)   Design extension event 1 (DEE-1): — limited damage.

(c)   Design extension event 2 (DEE-2): — severe damage (but no collapse).

The first level of structural performance — no damage (quasi-elastic behaviour) — is equivalent to the required structural performance under any other DBE (e.g. a design basis earthquake).

For design extension events (DEEs), structural damage is accepted. Two levels of performance are defined, depending on whether the containment safety function is preserved or not. Reinforced or pre-stressed structures are not leak tight. To achieve leak tightness, they need to be lined with a steel liner. The structural damage, in terms of support rotations as well as concrete scabbing, needs to be limited in order to prevent liner tearing and to preserve leak tightness. The function of the liner, other than for leak tightness, is to confine the scabbed concrete and to protect safety relevant SSCs.

5.1.2.   Subsystems

All safety relevant systems and components, even if protected by properly designed structures, need to be qualified/checked regarding their stability and functionality, considering the vibrations transferred from the impact locations to the region of their installation. In order to meet safety requirements, this qualification/checking needs to be performed for the whole frequency range and acceleration level of the corresponding response spectra at anchoring points. For many types of systems and components (especially instrumentation and control components), qualification is generally carried out by means of tests on shaking tables.

The stability and stress levels of safety relevant mechanical systems are usually assessed by means of analytical procedures, based on appropriate mathematical models and using the excitation (response spectra) at supporting points. This procedure results in a large amount of work, especially in cases of large, extended systems (e.g. piping) that have to be analysed for the entire frequency band between 0 and 80 Hz. In order to reduce the tremendous calculation effort, simplified design procedures have been established and introduced since the beginning of the 1980s in the design and layout of components and systems in Germany [72].

In this respect, it can be said that even if the aircraft crash or blast induced spectral accelerations are high, the associated spectral displacement values, beginning with a certain frequency range, could be very low (less than 1 mm) and, therefore, practically negligible. Displacements of such an order of magnitude are usually absorbed by elasto-plastic deformations between the component and its support, that is, without exhibiting significant stresses on the component [69]. The component design could, therefore, be performed based on a part of the acceleration response spectra (cut-off) between zero and the relevant frequency f_R or on the basis of a moderate, representative acceleration level b_R assumed to be constant within the whole frequency range. For example, in design practice, a constant (quasi-static) acceleration value may be defined for aircraft crash design.

5.2. ACCEPTANCE CRITERIA

5.2.1. General acceptance criteria

During DBEs, the three basic safety functions of a nuclear power plant (i.e. safe shutdown, decay heat removal, and containment and confinement) need to be ensured by maintaining structures in the elastic range (i.e. mostly recoverable deformation).

A DEE-1 represents events that would cause unrecoverable deformation (non-linearity) within structures but ensure the performance of all basic safety functions.

A DEE-2 includes those events that would cause damage to structures, leading to loss of confinement and containment, but would maintain the other two basic safety functions.

The events may be categorized into one of the above categories and the performance requirements may then be specified in terms of safety functions to be preserved for a given category of event, as presented in Table 12.

TABLE 12. EVENT LEVELS FOR SAFETY ASSESSMENT

Event level	Civil structure	Safety functions	Number of shutdown paths	Number of decay heat removal paths	Capacity assessment
Design basis event	Essentially elastic	Safe shutdown, decay heat removal, containment	Multiple	Multiple	Conservative
Design extension event 1	Plastic	Safe shutdown, decay heat removal, containment	2	2	Best estimate (median)
Design extension event 2	Plastic	Safe shutdown, decay heat removal	1	1	Best estimate (median)

5.3. SAFETY ASSESSMENT AGAINST MECHANICAL IMPACT

5.3.1. Failure modes

The failure modes to be addressed are:

— Overall stability;
— Overall structural capacity;
— Induced vibrations;
— Local failure modes.

5.3.2. Assessment of global failure modes

5.3.2.1. Overall stability

The global stability check is to be performed for the specified impact load acting on the upper regions of a building, taking into account the local soil conditions.

Global stability depends mainly on the overturning moment resulting from force acting on the highest impact location. Using the three dimensional model generated for the calculation of dynamic response analyses, the time histories of the resultant forces and moments acting on the foundation level can be derived for the representative impact location.

On the basis of the maximum forces and overturning moments acting in both horizontal directions, the safety against overturning of the building may be assessed considering the size and extension of uplift: the acceptance criteria are less than 30% uplift area at the base, with appropriate limitation of the compressive stresses in the foundation soil. The normal soil compressive stresses σ can be derived from the overturning moments and vertical loads acting at the base.

Global sliding also needs to be checked. Knowing the position of the neutral axis at the base, the sliding surface A and the torsional moment of inertia W_t of the sliding surface at the foundation mat can be obtained. Based on these results as well as on the horizontal forces F_H and torsion moment M_t obtained by dynamic analyses for the foundation level, the existing shear stresses can be derived as:

$$\tau_H = \frac{F_H}{A} \text{ and } \tau_T = \frac{M_\tau}{W_\tau} \tag{64}$$

Total shear stress is $\tau = \tau_H + \tau_T$. If σ is the average normal stress over the sliding surface at the base, it can be assumed that global sliding will not occur as long as the point (σ, τ) is located in region 1 of Fig. 33 [73]. It should also be taken into consideration that the embedment ratio of the building usually reduces the uplift and the risk of sliding.

In cases when the risk of sliding or overturning cannot be ensured against using these simplified stability checks, additional analyses are to be performed using a complex three dimensional mathematical model of the building and considering the surrounding soil.

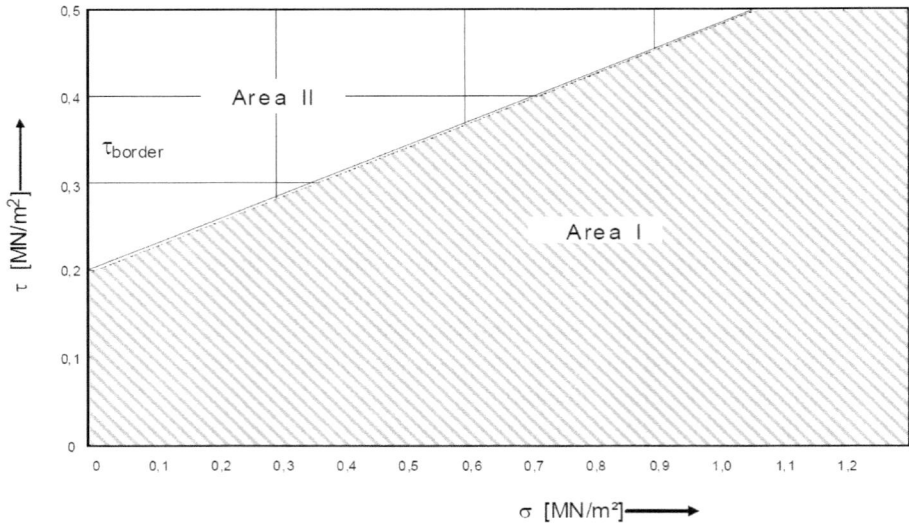

Borderline for $\sigma \geq 1.2$ MN/m^2: $\tau = 0.5$ MN/m^2

FIG. 33. Shear stresses and related average normal stresses during dynamic surface loading [73] (courtesy of the German Nuclear Safety Standards Commission).

5.3.2.2. Overall structural capacity

Overall structural capacity refers to the ability of the building structure to convey the impact loads to the foundation. In a containment type building (Fig. 34), this is done mainly by overall bending and shear.

Assessing overall structural capacity typically involves comparing strains, stresses, internal forces or moments computed during the structural response analysis with ultimate capacities. As described in Section 4, the structural response is commonly obtained from the explicit integration of finite element analyses considering material and geometrical non-linearities. Acceptance criteria then need to be in accordance with the finite element discretization, material models and specific parameters used in the computations.

When a segregated representation of steel and concrete is made (e.g. concrete and steel reinforcing bars are modelled using separate finite elements), acceptance criteria can be based on strain limits for each material. Table 13 provides guidance about the limits that can be used for the different levels of structural performance defined above. In this table, TF represents the 'triaxiality factor' of the stress state, which is defined as:

TABLE 13. STRUCTURAL ACCEPTANCE CRITERIA — ALLOWABLE STRAINS

Material	Strain measure	Allowable value for a design basis event	Allowable value for a design extension event 1 [8]
Carbon steel plate	Membrane principal strain (tensile)	0.010	0.050
	Local ductile tearing effective strain	n.a.[a]	0.140/TF[b]
304 stainless steel plate	Membrane principal strain (tensile)	0.010	0.067
	Local ductile tearing effective strain	n.a.[a]	0.275/TF[b]
Grade 60 reinforcing steel	Tensile strain	0.010	0.050
Post-tensioning steel (ungrouted tendons)	Tensile strain	0.010	0.030
Post-tensioning steel (grouted tendons)	Tensile strain	0.010	0.020
Concrete (for specified 28 d cylinder compressive strength less than 50 MPa)	Principal strain (compressive)	0.0035	0.0050[c]
	Principal strain (tensile)	d	d

[a] n.a.: not applicable.

[b] TF: triaxiality factor.

[c] The selection of an appropriate strain limit that defines concrete failure depends on the controlling type of failure — flexural or shear failure — and the biaxial or triaxial confinement conditions. The value given here corresponds to low confinement.

[d] Tensile cracking of the concrete is acceptable as long as the concrete tensile forces are transferred to the reinforcing steel, the steel does not exceed the allowable strain and the concrete does not exceed the compressive allowable strain. The concrete material model used in the computations needs to account for the reduction in shear capacity across cracks. Shear capacity is usually negligible for tensile strains larger than ten times the fracture strain. Fracture strain can be obtained by dividing the tensile strength by the initial Young's modulus.

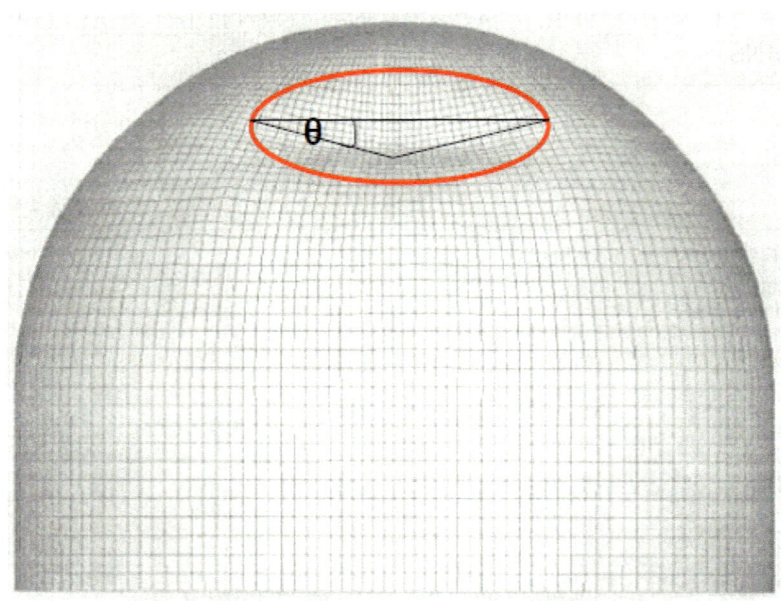

FIG. 34. A reactor building under a soft missile impact. Global behaviour — line of inflection [77].

$$TF = \frac{\sigma_1 + \sigma_2 + \sigma_3}{\sigma_{eq}} \qquad (65)$$

where σ_1, σ_2 and σ_3 are principal stresses, and σ_{eq} is the von Mises stress, also known as 'effective' or 'equivalent' stress. Conservatively, the TF can be taken as equal to two.

The values in Table 13 correspond to total deformation, including strains corresponding not only to impact loading, but also to other loads acting simultaneously (e.g. gravity loads).

The values given in Table 13 for DBE performance are basically the same as those used to compute structural capacity for other design loads, such as gravity loads. It should be noted that DBE criteria correspond to design basis cases, where essentially elastic structural behaviour, with irrelevant damage to the structure, is expected.

DEE-1 criteria correspond to the first tier of DEEs, where moderate structural damage is accepted. The structural behaviour can go into the plastic range, but with limited permanent deformations. Scabbing should either not occur or be limited. The presence of shear reinforcement significantly reduces the scabbing area and the scabbing is limited to the concrete cover [51, 52]. If the reinforced concrete structural element is lined with a steel liner, the liner should follow the deformation of the reinforced concrete element without rupture and the structure should stay leak tight. The scabbed concrete, if scabbing takes place, should be retained and confined by the liner.

DEE-2 criteria correspond to the second tier of DEEs, where there is significant damage but without structural collapse. At this performance level, a reinforced concrete structural element is seriously damaged but is still in place. Within a containment structure, there is scabbing of concrete mainly retained by the liner but the liner is breached and the structure is no longer leak tight. Strains for DEE-2 criteria are not given in Table 13, since beyond the levels indicated for DEE-1 the interpretation of the results from the analyses are very dependent on the computer code, the mesh and the material model.

When less sophisticated models are used to compute the structural response, strains of the different materials might not be available to use the acceptance criteria given in Table 13. In fact, historically, the first engineering acceptance criteria were derived to be used in the context of equivalent SDOF systems [9, 43]. For those systems, the permissible ductility ratio μ_a is defined as the ratio of the maximum acceptable displacement to the displacement at the effective yield point of the system. Ductility ratios are an indication of how many times the elastic strains at yield can be exceeded without reaching failure. Permissible ductility ratios depend on the stress state controlling the SDOF displacement (e.g. flexure, shear and compression) and they define the inelastic energy absorption capability of the equivalent SDOF system. This concept is still intensively used today for design against impact and explosion, not only in the nuclear industry [78, 79]. Table 14 provides permissible ductility ratios for reinforced concrete elements adapted from ACI 349-06 [42] and TM 5-1300 [9].

In an intermediate level of sophistication, shell or beam element models are used to compute the global structural response. In this case, generalized allowable strains can be defined in terms of permissible rotational capacities, θ_a at plastic hinges. Support rotation criteria are applicable both to SDOF equivalent systems and finite element models. The assessment of the rotation angle in finite element models, applied to a double-curvature dome, is illustrated in Fig. 33.

Table 14 includes permissible rotational capacities adapted from Refs [9, 42]. It should be noted that for a plastic hinge mechanism to be the governing failure mode, other more brittle modes, such as shear, compressive or buckling modes, need to be precluded. In the concrete structures normally present in nuclear facilities, it can be assumed that shear failure will not take place when plastic hinges develop before reaching 80% of the shear capacity at any cross-section. Table 14 is basically the same as that included in appendix I of Ref. [77].

TABLE 14. STRUCTURAL ACCEPTANCE CRITERIA FOR REINFORCED CONCRETE ELEMENTS *(adapted from Refs [9, 42])*

Element type	Design basis event	Controlling stress mechanism	Ductility ratio $\mu\alpha$	Flexure rotation in degrees[a,b] $\theta\alpha$	
			DEE-1, DEE-2	DEE-1	DEE-2
Beams	Essentially elastic behaviour[d]	Flexure Shear: concrete only concrete and stirrups stirrups only Compression	n.a.[c] 1.3 1.6 3.0 1.3	2	3
Slabs		Flexure Shear: concrete only concrete and stirrups stirrups only Compression	n.a.[c] 1.3 1.6 3.0 1.3	4	6
Beam/columns Walls Slabs in compression		Flexure Compression	1.3[e] 1.3	2	2
Shear walls Diaphragms		Flexure In-plane shear	n.a.[c] 1.5	1.5	2

[a] Transverse (shear) reinforcement is required for rotations greater than 2°.

[b] These rotation criteria (in degrees) are, in general, consistent with those in Ref. [42], which does not specify allowable inelastic deformation in terms of ductility ratio criteria for flexure. Figure 33 illustrates the concept of rotation angle.

[c] n.a.: not applicable.

[d] Essentially elastic behaviour means elastic structural analysis using design strain acceptance criteria of 1% for reinforcement in tension and 0.35% for concrete in compression. The permissible ductility ratio $\mu\alpha$ is 1.0.

[e] For additional detailed criteria, see section F.3.8 of Ref. [42].

106

In summary, for typical applications, Tables 13 and 14 present a tiered approach regarding acceptance criteria with respect to the three levels of overall structural performance defined above (DBE, DEE-1 and DEE-2). The values in Tables 13 and 14 are maximum values under the loading condition being considered, which need to include all loads acting simultaneously.

The analyst should be conscious that appropriate structural detailing is required to achieve the allowable values given in the tables and that the computational models usually do not include a full representation of the details. Hence, detailing rules for impact loading given in concrete [42] or structural steel [78] standards need to be respected. For example, for DEE-2 performance of a reinforced concrete slab, the longitudinal reinforcement has to be adequately developed into supports if credit is given to catenary action.

5.3.2.3. Induced vibrations

Subsystems (e.g. safety systems, components, equipment and distribution systems) are subject to two types of acceptance criteria, corresponding to two intended functions: structural integrity and functionality. The structural integrity of ductile SSCs is generally unaffected by high frequency input motions. However, the functionality of some devices, such as relays, can be affected by low and high frequency input motions.

To assess structural integrity, a truncated in-structure response spectrum approach is selected. Displacement response spectra are calculated and the frequency at which the spectral displacement reaches a threshold value, such as 1 mm, is defined. For frequencies equal to or greater than this threshold frequency, the acceleration response spectra are truncated and the spectral acceleration value at this frequency is extended to all high frequencies of interest. Assessments of the capacity of the subsystem are performed according to the following criteria:

(a)   Comparing the truncated in-structure response spectra with the in-structure response spectra used for other DBEs and DEEs (e.g. seismic, hydrodynamic loads and aircraft crash). If the in-structure response spectra for other DBEs or DEEs envelop the truncated in-structure response spectra, and the performance criteria of the subsystem are the same as for the DBEs and DEEs of interest, the subsystem is considered to have met the performance criteria.

(b)   If the truncated in-structure response spectra are not enveloped by the in-structure response spectra used for other DBEs and DEEs, then analysis, testing or other qualification methods are required. Innovative techniques supported by adequate justification may be implemented to demonstrate the capacity of the subsystem to achieve its performance requirements.

To assess functionality related requirements, the following steps may be followed:

(a)   Identifying important subsystems the performance of which is susceptible to high frequency input motions (e.g. some relays, switches and controllers).
(b)   Assessing whether these subsystems will be adversely affected by chatter (or position change) or whether the subsystem can be easily recoverable through automatic or operator action.
(c)   If it is important, assessing the consequences of the high frequency input motions on subsystem performance and defining the loading environment to be represented by the non-truncated in-structure response spectra or other generic parameters, such as acceleration values used in testing components.
(d)   Seeking existing test data applicable to these subsystems and comparing the loading environment to the input motions from existing test data:
      (i) If the existing test data envelop the loading environment and the existing data are for the same functional performance requirements, the subsystem is considered to meet the criteria;
      (ii) If not, testing the subsystem or replacing the subsystem if it is susceptible to failing to perform its function when subjected to high frequency motion may be required.

5.3.3.   Local failure modes

Local failure modes are those taking place in the immediate vicinity of the impact point. Local effects consist of [42]:

— Penetration: displacement of a missile into an impacted structural element.
— Perforation: the passing of a missile completely through the impacted structural element, with or without exit velocity.
— Punching shear: local shear failure occurring in the immediate vicinity of the impacted zone. A punching shear failure occurs as part of a perforation.
— Scabbing: ejection of material from the back face of the impacted structural element opposite the face of impact.
— Spalling: ejection of material from the front face of the impacted structural element, that is, the face on which the missile impacts.

The local response of the target will be initiated with spalling and subsequent penetration of the target. If the missile has enough energy, at some stage, scabbing will take place and, eventually, perforation (Fig. 35 [80]).

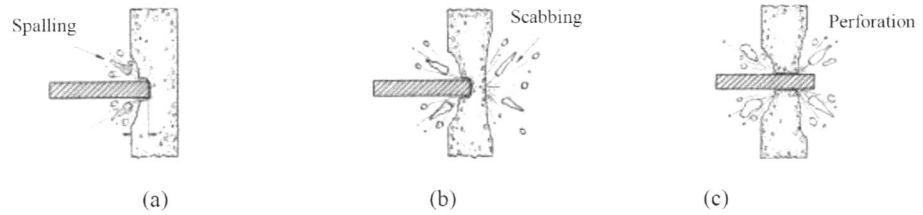

FIG. 35. Missile impact phenomena: (a) missile penetration and spalling; (b) target scabbing; (c) perforation [80].

It should be noted that these definitions are not universally used; 'back face spalling' is sometimes used instead of scabbing to refer to the ejection of materials from the back face.

Empirical formulas validated by tests may be used to predict these local responses for predominantly rigid (non-deformable) missiles. Empirical formulas are used for cases of normal (90°) impact. When the impacting missile strikes the target face normally, the local responses are maximized. The angle of the strike can substantially influence the extent of local damage and should be taken into consideration [81].

In assessing the ultimate capacity of the target, acceptance criteria also need to be chosen for these local failure modes. It is worth noting that the yield strength of the materials can be increased due to the high strain rates taking place. However, this is not necessarily the case with the ultimate strains, because fracture energy does not increase. On the contrary, for steel, the ultimate elongation corresponding to the elevated tensile strength is lower than that in a static case. In other words, the material is more brittle at high strain rates.

5.3.3.1. Punching shear

The static punching capacity of a concrete slab with no transverse reinforcement (i.e. no stirrups) can be obtained from the formula:

$$F_{\mathrm{p}} = 8170 \left(\rho_{\mathrm{p}} f_{\mathrm{c}} \right)^{1/3} \pi d_{\mathrm{e}} \left(d_{\mathrm{load}} + 2.5 d_{\mathrm{e}} \right) \tag{66}$$

where

ρ_{p} is the average ratio of tensile steel reinforcement on the tensioned face (%);
f_{c} is the compressive strength of concrete (Pa);
d_{e} is the distance between the front face and reinforcement (m);

and d_{load} is the diameter of the loaded area (m) [82].

Equation (66) is experimentally verified in the range of:

$$0.07 \text{ m} < d_e < 0.9 \text{ m}$$

$$0.66 < d_{load}/d_e < 1.3$$

$$0.22\% < \rho_p < 1.26\%$$

$$25 \text{ MPa} < f_c < 63 \text{ MPa}$$

$$0.05 < a_g/d < 0.07$$

where a_g is the aggregate size of concrete.

According to Ref. [82], the static punching shear resistance formula can be applied for dynamic soft impact cases by checking the condition:

$$\bar{F} \leq F_p \tag{67}$$

where $\bar{F}$ is the average value of the time dependent force resultant of the missile, which can be calculated as:

$$\bar{F} = \frac{0.9I}{t_{0.9I}} \tag{68}$$

where $t_{0.9I}$ is the time when 90% of the total impulse $(0.9I)$ is reached during the dynamic loading transient. This means, in practice, that the possible long tail of the loading function $F(t)$ is discarded.

The required wall thickness can be estimated, based on the concrete quality, mass and velocity of the missile, using the following empirical formula [83]:

$$d = \left(\frac{3}{10^4 \beta_{WN}^{1/2}}\right)\left(\frac{M}{D^{1.8}}\right)\left(\frac{v}{100}\right)^{4/3} \tag{69}$$

where

d is the minimum required thickness (cm);
M is the mass of the engine missile (kg);
D is the diameter of the engine missile (cm);

v is the impact velocity of the engine (m/s);

and β_{WN} is the nominal ultimate compressive strength of concrete (kg/cm^2).

5.3.3.2. Scabbing thickness

Empirical formulas for scabbing thickness have been developed by several investigators and research institutes. The formulas give the required minimum wall thickness to avoid scabbing in a concrete target. The existing empirical formulas are given below, with the following symbols:

D is the effective missile diameter (equivalent diameter of the contact area) (m);
$f_c{'}$ is the concrete compressive strength (MPa);
t_s is the scabbing thickness (m);
t_p is the perforation thickness (m);
M is the mass of the missile (kg);
W is the weight of the missile (N);
v is the missile impact velocity (m/s);
x is the penetration depth of the missile (m);
N is the missile shape factor;
K is the concrete penetrability factor.

Typically, empirical formulas are based on data for lightly reinforced (0.3% to 1.5% each way) concrete targets, with no transverse reinforcement (i.e. no stirrups). Application to heavily reinforced targets would give a conservative estimate.

It should be noted that most formulas were developed using tests with solid projectiles. When used for aircraft engine parts or landing gear, reduction factors between 0.60 and 0.65 are usually appropriate to take into account the deformability of the projectile [84].

In addition, it should be noted that the tests performed under the IMPACT programme showed that the presence of transverse reinforcement significantly reduces the scabbing area [51, 52]. It also reduces the volume of the scabbed concrete to the concrete cover.

Chang formula

The empirical formula proposed in 1981 in Ref. [85], based on lower velocity missiles (less than 150 m/s), to predict the scabbing thickness for reinforced concrete panels subjected to a cylindrical rigid (non-deformable) steel missile impact:

$$t_s = \frac{0.005 W^{0.4} v^{0.67}}{D^{0.2} f_c'^{0.4}} \ \text{(m)} \tag{70}$$

The application limits are [57]:

16 m/s $\leq v \leq$ 312 m/s

22.8 MPa $\leq f_c' \leq$ 45.5 MPa

1.08 N $\leq W \leq$ 3365 N (0.11 kg $\leq M \leq$ 343 kg)

0.0505 m $\leq D \leq$ 0.305 m

To prevent scabbing, a safety factor of 1.1 is usually applied to t_s [80].

Modified Chang formula

The modified Chang formula for the minimum wall thickness to prevent scabbing is presented in Refs [8, 86]. It corresponds to the Chang formula multiplied by a reduction factor a_s. In SI units, the equation can be given as:

$$t_s = \alpha_s 1.84 \left(\frac{u}{v}\right)^{0.13} \left(\frac{M v^2}{D^{0.5} f_c'}\right)^{0.4} 0.004 \ \text{(m)} \tag{71}$$

where u is the reference velocity; u = 61 m/s. The recommended value for α_s is 0.55.

Stone and Webster formula

Based on a series of one quarter scale tests, Stone and Webster developed an empirical formula [57] for predicting scabbing thickness of concrete targets struck by solid steel and pipe missiles with velocities typical for nuclear plant applications. In the case of a solid steel missile, the formula for scabbing thickness is given as:

$$\frac{t_s}{D} = \left(\frac{W v^2}{23.8 \times 10^6 D^3}\right)^{\frac{1}{3}} \tag{72}$$

The application limits are [57]:

23 m/s $\leq v \leq 76$ m/s

20.7 MPa $\leq f_c' \leq 31$ MPa

$1.5 \leq t_s/D \leq 3.0$

The formula is based on tests with rather thin concrete plates, between 10 and 15 cm thick [87].

Central Research Institute of the Electric Power Industry formula

The Central Research Institute of the Electric Power Industry (CRIEPI) of Japan conducted impact testing focusing on low velocity missiles and proposed the CRIEPI formula [57, 88]. This formula amended the Chang formula. According to this formula, the scabbing thickness is given by:

$$t_s = \frac{0.0047W^{0.4}v^{0.67}}{D^{0.2}f_c'^{0.4}} \ \text{(m)} \tag{73}$$

It should be noted that the formula gives a scabbing thickness 95% of that given by Eq. (70).

5.3.3.3. Perforation thickness

Empirical formulas for perforation thickness have been developed by several investigators and research institutes. The perforation thickness t_p is defined as the concrete thickness that is just large enough to allow a missile to pass through the panel without any exit velocity [89]. The formulas usually give the mean empirical value. A safety coefficient of 1.2 normally needs to be applied to the results obtained from the formulas to cover test uncertainty. Additionally, it should be noted that most formulas were developed using tests with solid projectiles. When used for aircraft engine parts or landing gear, reduction factors of between 0.60 and 0.65 are usually appropriate to take into account the deformability of the projectile [84].

It should be noted that, when the thickness is given, the formulas below (Eqs (74–84)) can be used to obtain the perforation velocity of a particular missile.

As mentioned earlier, these empirical formulas are typically based on data for lightly reinforced (0.3% to 1.5% each way) concrete targets, with no transverse reinforcement (i.e. no stirrups). Application to heavily reinforced targets would yield a conservative estimate. In addition, it should be noted that the tests performed in the IMPACT programme showed that the influence of transverse reinforcement on perforation depends on its anchorage system but also that the influence of transverse reinforcement is limited [52].

The empirical formulas are given below (Eqs (75–85)). The symbols are the same as for the scabbing formulas. Punching shear considerations for small non-deformable missiles are implicit in the formulas.

French Alternative Energies and Atomic Energy Commission–Électricité de France formula

Based on an empirical fit to data from 52 tests with solid cylinder missiles impacting at velocities greater and less than the critical perforation velocity, the French Alternative Energies and Atomic Energy Commission–Électricité de France (CEA–EDF) formula [90, 91] was developed for perforation thickness in France in 1977. It gives the perforation thickness as:

$$t_p = 0.82 \frac{M^{0.5} v^{0.75}}{\rho^{0.125} D^{0.5} \left(f_c' \times 10^6 \right)^{0.375}} \text{ (m)} \tag{74}$$

where ρ is the density of the concrete (i.e. 2500 kg/m³) which is introduced as a pseudo-parameter for dimensionality considerations.

The application limits are [90]:

25 m/s $\leq v \leq$ 450 m/s

29.6 MPa $\leq f_c' \leq$ 50.3 MPa

149 kg/m³ $\leq$ steel reinforcement quantity (symmetrical) $\leq$ 298 kg/m³

0.349 $\leq t/D \leq$ 4.17 (t = slab thickness)

Chang formula

In 1981 Chang [85] also proposed an empirical formula, that was similar to the scabbing thickness formula, based on lower velocity missiles (less than 150 m/s) to predict the perforation thickness for reinforced concrete panels subjected to a cylindrical, rigid (non-deformable) steel missile impact:

$$t_\mathrm{p} = \frac{0.9\times 10^{-3}\,W^{0.5}v^{0.75}}{D^{0.5}f_\mathrm{c}'^{\,0.5}}\ \mathrm{(m)} \tag{75}$$

The application limits are [57]:

16 m/s $\leq v \leq$ 312 m/s

22.8 MPa $\leq f_\mathrm{c}' \leq$ 45.5 MPa

1.1 N $\leq W \leq$ 3430 N (0.11 kg $\leq M \leq$ 343 kg)

0.051 m $\leq D \leq$ 0.305 m

Central Research Institute of the Electric Power Industry formula

CRIEPI [57] of Japan conducted impact testing focusing on low velocity missiles and proposed the CRIEPI formula, which amended the Chang formula. In the CRIEPI formula the perforation thickness is given by:

$$t_\mathrm{p} = \frac{0.8\times 10^{-3}\,W^{0.5}v^{0.75}}{D^{0.5}f_\mathrm{c}'^{\,0.5}}\ \mathrm{(m)} \tag{76}$$

It should be noted that the formula gives a scabbing thickness 90% of that given by Eq. (75).

National Defence Research Committee formula

The National Defence Research Committee (NDRC) introduced the following formula for the calculation of penetration depth x:

$$K\cdot N\cdot M\frac{1}{D}\left(\frac{v}{59\,525 D}\right)^{1.8} = \begin{cases} \left(\dfrac{x}{2D}\right)^2 & \text{when } \left(\dfrac{x}{D}\right)\leq 2 \\[2mm] \dfrac{x}{D}-1 & \text{when } \left(\dfrac{x}{D}\right)\geq 2 \end{cases} \tag{77}$$

where

N is a missile nose shape factor:

— Flat nose: $N = 0.72$.
— Blunt nose: $N = 0.80$.
— Hemispherical nose: $N = 1.00$.
— Very sharp nose: $N = 1.14$.

and K is a concrete penetrability factor which is a function of concrete strength.

The NDRC effort was stopped without completely defining the factor K. The parameter K in the original NDRC formula was later replaced by [81]:

$$K = \frac{15}{\sqrt{f_c'}}$$

The application limits are [92]:

25 m/s $\leq v \leq$ 300 m/s

22 MPa $\leq f_c' \leq$ 44 MPa

5000 kg/m$^3 \leq M/D^3 \leq$ 200 000 kg/m^3

Modified National Defence Research Committee formula

The modified NDRC formula [81] gives the penetration depth x using the empirical equations given in Eq. (77):

$$x = \sqrt{4K \cdot N \cdot M \cdot D \left(\frac{v}{59\ 525D}\right)^{1.8}} \qquad \text{when } \frac{x}{D} \leq 2 \text{ (m)}$$

$$x = K \cdot N \cdot M \left(\frac{v}{59\ 525D}\right)^{1.8} + D \qquad \text{when } \frac{x}{D} > 2 \text{ (m)}$$

(78)

Once x is known, the perforation thickness t_p is given by:

$$\frac{t_p}{D} = 1.32 + 1.24\left(\frac{x}{D}\right) \qquad \text{for } 1.35 < \frac{x}{D} < 13.5$$

$$\frac{t_p}{D} = 3.19\left(\frac{x}{D}\right) - 0.718\left(\frac{x}{D}\right)^2 \qquad \text{for } \frac{x}{D} \leq 1.35$$

(79)

The application limits are [93]:

25 m/s $\leq v \leq$ 300 m/s

22 MPa $\leq f_c' \leq$ 44 MPa

5000 kg/m$^3 \leq W/D^3 \leq$ 200 000 kg/m^3

Degen formula

A partial revision of the modified NDRC formula was developed by Peter Degen to predict perforation thickness [94]. Based on the analysis of experimental results in Ref. [94], new coefficients have been calculated for the above NDRC perforation thickness formulas in Eq. (79):

$$\frac{t_p}{D} = 0.69 + 1.29\left(\frac{x}{D}\right) \qquad \text{for } 2.65 < t_p/D < 18 \text{ or } 1.52 < x/D < 13.42$$

(80)

$$\frac{t_p}{D} = 2.2\left(\frac{x}{D}\right) - 0.3\left(\frac{x}{D}\right)^2 \quad \text{for } t_p/D < 2.65 \text{ or } x/D < 1.52$$

The penetration depth x is obtained from the NDRC formulas in Eq. (78). The application limits are [57]:

25 m/s $\leq$ v $\leq$ 312 m/s

28.4 MPa $\leq f_c' \leq$ 43.1 MPa

159 kg/m^3 $\leq$ steel reinforcement quantity $\leq$ 348 kg/m^3

0.10 m $\leq D \leq$ 0.31 m

0.15 m $\leq t \leq$ 3.0 m (t = slab thickness)

When the Degen formula is applied to the impact of aircraft engines or landing gear, the deformability of the projectile can be taken into account using a reduction factor when estimating the perforation thickness t_p [84]:

$$\frac{t_p}{D} = \alpha_p\left[2.2\left(\frac{x}{\alpha_p D}\right) - 0.3\left(\frac{x}{\alpha_p D}\right)^2\right] \qquad \text{for } x/\alpha_p D < 1.52$$

(81)

The recommended value for a_p is 0.60 [86].

United Kingdom Atomic Energy Authority formulas

The United Kingdom Atomic Energy Authority (UKAEA) proposed the following formulas for calculating the penetration depth x [89]:

$$\frac{x}{D} = 0.275 - \sqrt{0.0756 - G} , \text{ when } G < 0.0726 \tag{82}$$

or

$$\frac{x}{D} = \sqrt{4G - 0.242} , \text{ when } 0.0726 \le G \le 1.0605$$

or

$$\frac{x}{D} = G + 0.9395 , \text{ when } G > 1.0605$$

where the function G is the same as in the NDRC formulation:

$$G = K \cdot N \cdot M \frac{1}{D} \left(\frac{\upsilon}{59\,525 D} \right)^{1.8}$$

The application limits are [57]:

25 m/s $\le \upsilon \le$ 300 m/s

22 MPa $\le f_c' \le$ 44 MPa

5000 kg/m^3 $\le W/D^3 \le$ 200 000 kg/m^3

In addition, the UKAEA gave the following formulas for obtaining missile velocity which results only in perforation υ_p. These formulas are an elaboration of previous work by the CEA–EDF [84].

$$\upsilon_p = \upsilon_a \text{ (m/s) } \quad \text{for } \upsilon_a \le 70 \text{ m/s}$$

$$\tag{83}$$

$$\upsilon_p = \upsilon_a \left[1 + \left(\frac{\upsilon_a}{500} \right)^2 \right] \text{ (m/s) } \quad \text{for } \upsilon_a > 70 \text{ m/s}$$

where

$$\upsilon_a = 1.3 \rho_c^{1/6} \sqrt{k_c \times 10^6} \left(\frac{p \cdot t^2}{\pi M} \right)^{\frac{2}{3}} \sqrt{\rho_p + 0.3 \left[1.2 - 0.6 \left(\frac{c_r}{t} \right) \right]} \text{ (m/s);}$$

ρ_c is the density of concrete (kg/m^3);
ρ_p is the amount of reinforcement (%) (each face, each way);

t is the plate thickness (m);

p is the perimeter of the missile cross-sectional area (m);

and c_r is the rebar spacing (m) [89]. The parameter k_c is equal to the unconfined compressive strength of the concrete of the target f_c' (MPa), when it is less than 37 MPa, and equal to 37 MPa, when it exceeds this value.

For targets with different reinforcement amounts on the front and rear faces, the parameter ρ_p is the sum of one third times the 'each way' value on the front face and two thirds times the 'each way' value on the rear face.

The application limits are [84, 89]:

11 m/s $\leq v \leq$ 300 m/s

22 MPa $\leq f_c' \leq$ 52 MPa

$0.0 \leq \rho_p \leq 0.75\%$ (each way, each face)

$0.2 \leq p/(\pi t) \leq 3$

150 kg/m^3 $\leq M/(p^2 t) \leq$ 10 000 kg/m^3

$0.12 \leq c_r/t \leq 0.49$ for $0.49 \leq c_r/t$, assume $1.2 - 0.6\left(\dfrac{c_r}{t}\right) = 1.0$

University of Manchester, Institute of Technology (UMIST) formula

The following penetration depth formula was developed within a research programme by UK Nuclear Electric at the University of Manchester, Institute of Technology [57]:

$$\frac{x}{D} = \left(\frac{2}{\pi}\right)\frac{N}{0.72}\left(\frac{Mv^2}{\sigma_t D^3}\right) \tag{84}$$

where the nose shape factor N is 0.72 for a flat nose, 0.84 for a hemispherical nose, 1.0 for a blunt nose and 1.13 for a sharp nose, and:

$$\sigma_t = 4.2f_c' + 135 \times 10^6 + \left(0.014f_c' + 0.45 \times 10^6\right)v \text{ (Pa)} \tag{85}$$

is a rate dependent strength parameter of concrete, for which the f_c' values should be introduced in pascals and the missile velocity v, in metres per second.

The application limits are [57]:

$$3 \text{ m/s} \le v \le 66.2 \text{ m/s}$$

$$0.05 \text{ m} \le v \le 6.00 \text{ m}$$

$$35 \text{ kg} \le M \le 2500 \text{ kg}$$

$$0 \le x/D \le 2.5$$

Comparison between formulas

The application limits and validity of the formulas are verified in the following against test results. The penetration depth of a test slab impacted by a hard missile was predicted by various formulas. The characteristics of the rigid missile are: diameter $D = 0.17$ m and mass $M = 47$ kg. The thickness of the target plate is $t = 0.25$ m and the compressive cube strength of concrete is 54 MPa. The cylinder strength can be estimated to be $f_c' = 45.9$ MPa.

The penetration depth x was calculated as a function of the impact velocity and the results obtained using various formulas are presented in Fig. 36, together with some test results. These tests were carried out with an impact velocity v of 100 m/s. The reinforcement of the test slabs was varied in order to study the effect of shear reinforcement and pre-stressing [95].

It should be noted that the presence of the T-bars reduces the penetration depth and that the differences between the values predicted by the formulas are significant. Application of the formulas requires a careful examination of the experimental background and the range of validity.

5.3.3.4. Residual velocity

If the calculated perforation thickness t_p is greater than the wall thickness t, then the wall is perforated by the missile, which comes out with a residual velocity v_r [96]:

$$v_r = \sqrt{\frac{v^2 - v_p^2}{1 + \dfrac{M_k}{M}}} \qquad (86)$$

where v is the initial missile velocity;

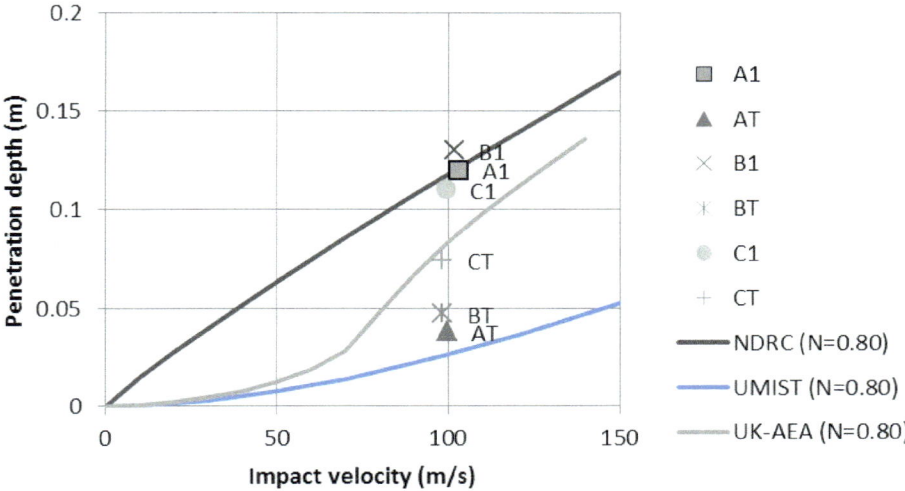

FIG. 36. Penetration depth of a test slab predicted by various formulas. The lower point set, referring to test results, at about 100 m/s, with T-bars included, is in the order AT, BT, CT from the bottom. The upper point set, at the same impact velocity, is in the order C1, A1, B1 from the bottom.

v_p is the 'just perforation' velocity (calculated with one of the formulas above);
M_k is the mass of the ejected concrete;

and M is the missile mass.

The mass of ejected concrete can be calculated using the conical plug geometry developed in Ref. [96], where the volume of the ejected concrete is that of a conical frustum, with minor radius $r_1 = D/2$, and major radius $r_2 = r_1 + t$ (tan θ), where $\theta = 45°/(t/D)^{1/3} \le 60°$.

5.4. STRUCTURAL SAFETY ASSESSMENT AGAINST EXPLOSION

5.4.1. Safety assessment against distant explosion

The structural acceptance criteria given in Tables 13 and 14 are also applicable in safety assessments carried out for distant explosions, taking into account that the loading will be of the impulsive type in those cases.

5.4.2. Safety assessment against contact and near field blasts

The thickness of elements subjected to contact blasts will be determined by local effects.

As mentioned above, scabbing or back face spalling is defined as the ejection of fragments of a structural element from the opposite side from which the structural element was loaded [97]. The stress wave induced by the load propagates through the material away from the loading source. When the stress wave reaches a free surface, it will reflect. During reflection, the front reflected portion of the stress wave combines with the back incident portion to form a net stress wave.

If the net stress exceeds the dynamic tensile strength of the material for a long enough time, the material will crack. The impulse of the portion of the stress wave trapped between the crack and the back free surface equals the momentum imparted to the cracked off portion in the direction away from the structural element. If the trapped impulse is great enough to overcome the resistive forces, the cracked off part separates from the backside of the structure at some velocity.

Resistive forces are due to bond shear and mechanical interlocking. The remaining portion of the stress wave will reflect off the newly formed free surface. This process continues until either the remaining stress wave cannot cause back face spall or the structural element is breached. Stress waves also change in magnitude and shape as they propagate through real material. These changes are due to several causes: different stresses propagate at different velocities; attenuation; divergence of the wave energy over an expanding wave; and dispersion of stress waves upon striking air voids, reinforcement steel and other imperfections.

Mills [98] presents simple rules for the minimum thickness T of a concrete slab to ensure that back face spalling is either prevented or limited:

— To resist back face spalling: $T/W^{1/3} > 0.32$ m/kg$^{1/3}$.
— To allow light back face spalling: $T/W^{1/3} > 0.27$ m/kg$^{1/3}$.
— To allow heavy back face spalling: $T/W^{1/3} > 0.23$ m/kg$^{1/3}$.

W is the TNT equivalent weight of the explosive.

Spall and breach limits for concrete walls can be determined based on experiments [97]. The damage classification of a concrete wall is defined as follows:

— No damage: from no change in the condition of the wall to a few barely visible cracks.

— Threshold spall: from a few cracks and a hollow sound to a large bulge in the concrete wall with a few small pieces on the floor.
— Breach: from a small hole which barely lets light through to a large hole.

Basler [99] presents empirical back face spall prediction curves based on 96 tests with bare charges from 0.014 to 227 kg. The stand-off distance varied from contact to 1.2 m. The wall thicknesses varied from 9 to 110 cm; the static compressive strengths of the concrete, from 20 to 55 MPa; and the reinforcement ratio from 0.64% to 1.91%. According to measurements in fig. 3.4 of Ref. [99], the spall limit is approximately:

$$y = 0.060x^{-0.565} \tag{87}$$

where

$$x = \frac{R}{W^{1/3}} \; ;$$

$$y = \frac{T}{W^{1/3}} \; ;$$

T is the slab thickness (m);
R is the stand-off distance of the explosive charge (m);

and W is the TNT equivalent weight of explosive charge (kg).

In the same way, the breach limit can be determined in the form:

$$y = 0.031x^{-0.55} \tag{88}$$

Experimental results and empirical damage curves are shown in Fig. 37.

McVay [97] conducted 40 spall tests to investigate the parameters affecting spall and to check existing prediction methods. The explosive sizes varied from 0.67 to 18.3 kg equivalent TNT weight and the stand-off distances from contact to 1.52 m. Tests were conducted on wall thicknesses of 13.6, 21.6 and 28.5 cm. The concrete compressive strength varied from 28 to 97 MPa.

The spall depths and spall velocities observed in tests on high strength concrete were much higher than in tests on ordinary 28 MPa concrete. In high strength concrete walls, stress waves behave more elastically and do not change magnitudes and shapes as much as in ordinary concrete. In addition, spall damage

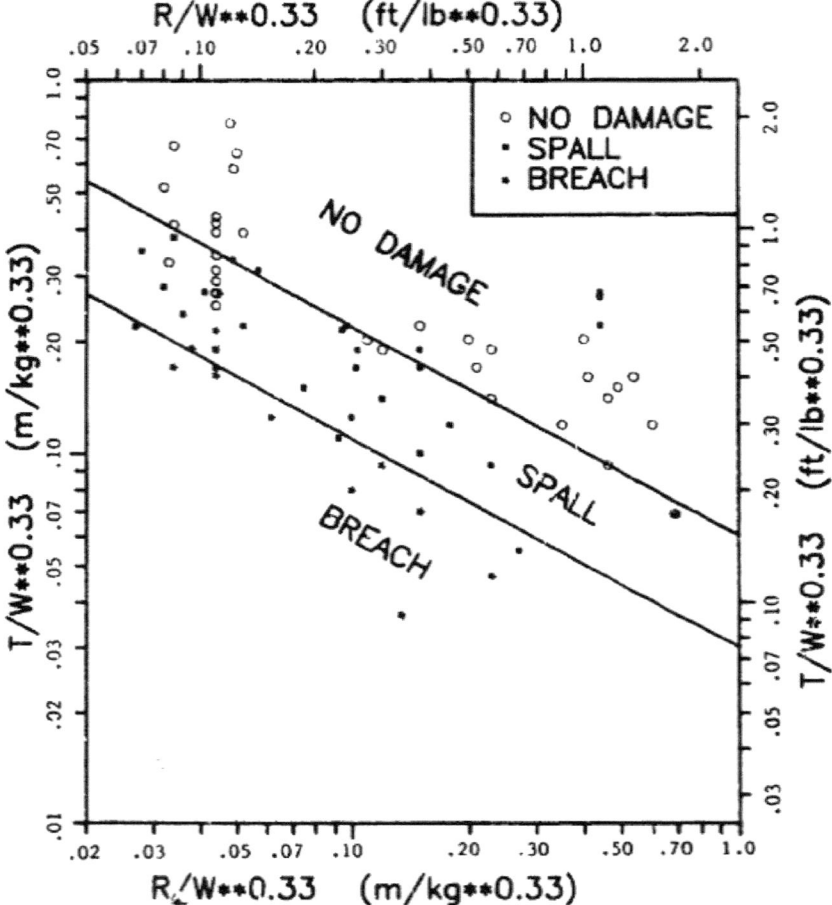

FIG. 37. *Empirical damage prediction curves from Refs [97, 99] for damage of reinforced concrete panels subjected to nearby detonations (** in the axis labels denotes exponentiation).*

is worse in large scale detonations than in small scale detonations [97]. The above presented test data are not conservative for high strength concrete.

The back face spall limit, based on experiments reported in Ref. [97], is:

$$y = 0.070x^{-0.6} \tag{89}$$

The breach limit is:

$$y = 0.032x^{-0.59} \tag{90}$$

124

The spall and breach limits calculated by these equations are shown in Fig. 38.

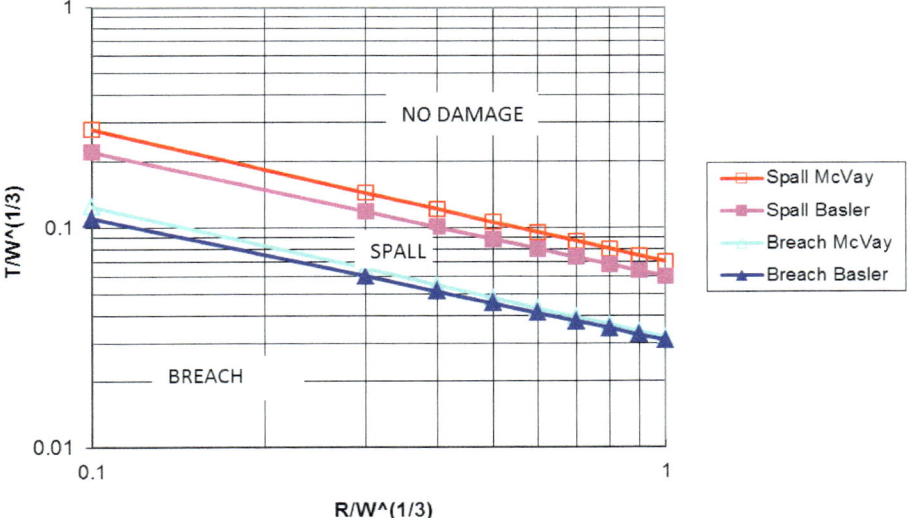

FIG. 38. Log-log prediction curves for damage to concrete panels caused by bare charges (adapted from Ref. [97]).

5.5. FIRE PERFORMANCE OF FIRE BARRIERS AND COMPONENTS

5.5.1. Fire barrier performance

5.5.1.1. General principles

The main principle for reducing the fire risk in a nuclear installation is based on the division of the buildings into fire compartments or fire areas using fire barriers [3]. The fire barriers consist of structural components with an appropriate fire resistance rating. To protect the nuclear installation from external fires, the external walls and roofs of the safety relevant buildings also need to be robust structures to prevent, for example, jet fuel from entering the building and, additionally, they need to have a sufficient fire resistance rating to withstand the external heat exposure. The fire resistance rating of a barrier component or structure is denoted as a number (e.g. 60) that gives the time in minutes the component has been able to withstand a standardized test. The rating can also indicate the criteria that were tested. In Europe, for instance, the three criteria are denoted as 'R' (load bearing), 'E' (integrity) and 'I' (insulation). The requirements and details of the test procedures are defined on a national basis. In addition, the fire barrier components need to be rated for a pressure difference to account for the overpressure generated by fireballs and vapour cloud explosions.

All of the components of the fire barrier have the same rating. Load bearing capacity is, of course, only required for the load bearing structures, such as walls and horizontal slabs (ceiling/floor). Examples of fire barrier components are fire doors, fire dampers and penetration seals for cables and ducts.

Most fire barrier components are passive by nature, that is, they are not expected to change their state in case of fire. The most significant group of active barrier components are the fire dampers which close during the fire, either by thermal or electrical activation.

5.5.1.2. Conservative screening approach for the fire performance of barriers

A conservative approach, developed for aircraft crash assessment [8], is to assume that the fire will propagate to an area surrounded by:

(a)   A single fire barrier rated for at least 3 h and a 30 kPa pressure difference; or

(b)   Two fire barriers rated for at least 3 h and less than a 30 kPa pressure difference.

This approach acts as a first level screen. It assumes that ventilation ductwork with a less than 30 kPa pressure difference rating will provide a pathway for fire product propagation due to the induced overpressure of the fire.

5.5.1.3. Simple engineering approach for the assessment of the fire performance of barriers

The performance of the structural components forming the barriers can be assessed by using fire safety engineering methods to estimate the severity and duration of the fire. Fire analysis needs to determine:

— Whether the fire will lead to flashover or whether it will remain local;
— The heat release rate of the fire;
— The temperature of the flashover fire or local fire;
— The fire duration in hours.

The methods for addressing these points can be found, for instance, in Refs [22, 40], which provide the necessary spreadsheet tools. Finally, the performance of the barrier can be assessed by comparing the fire duration to the fire rating (hours) of the structural components. If the rating is less than the estimated fire duration, the barrier will fail.

5.5.1.4. Detailed approach for fire barrier performance assessment

Computational analysis can be used as a detailed approach for fire barrier performance assessment. The performance of the barrier components is based on the temperature failure criterion or detailed mechanical analysis, as explained in Section 4.4.

Simple failure criteria can be based on the calculated temperatures. Concrete structures may experience adverse effects due to extreme heating rates. Owing to the capillary water present in concrete, explosive spalling may occur, which leads to rapid loss of concrete cover and the possibility of a direct fire attack of the reinforcing steel. The steel reinforcement bars lose their strength above 400°C and the whole structure may lose its load bearing function. Pre-stressed concrete members may also lose their load bearing function if pre-stressing cables are heated above 250°C.

Steel structures or components are also sensitive to fire exposure. Owing to their high thermal conductivity and heat absorption, steel failure may be assumed at 500°C for structural steel. The supports of large components may fail in the case of intensive fire exposure due to loss of strength of columns, hangers, ribs, supports, etc. early on in the fire scenario.

5.5.1.5. Fire damper performance

The capacity of the fire dampers to perform their barrier function depends on both their thermomechanical stability (capacity to maintain leak tightness and thermal insulation under rapid and non-uniform heating and pressure difference) and the response of the closing mechanism. As with any other barrier components, the fire dampers need to be rated for a sufficient length of time for a standard fire test and pressure difference.

A special feature of human induced external events, such as explosions and aircraft crashes, is the rapid time evolution of the exposure. Fast activation and response are needed for:

— Closing the air intake channels feeding fresh air to the various parts of the plant, e.g. main control room, electronics rooms and diesel engines;

— Closing the ventilation ducts outside the physical damage footprint to prevent the spreading of the fire.

The response time of the damper depends on the physical characteristics and closing mechanism. The reliability of the closing mechanism under a pressure wave may be difficult to verify. The margin between the damper closing and the fireball arrival times depends on the speed of the activation mechanism, the physical distance between the event and ventilation shaft entrance, and the length and shape of the shaft leading from the entrance to the damper. In aircraft crashes, for instance, the speed of the fireball expansion is about 40 m/s. Inside a ventilation shaft, the fireball will proceed with a speed that is lower than the free air value, although the air speed inside the shaft can be significantly higher. The activation of the damper is to be based on as early detection of the external event as possible. Optical methods that react to the conditions outside the plant are needed.

5.5.2. Fire performance of safety related structures, systems and components

5.5.2.1. Conservative screening approach

A conservative approach is to assume that structures and systems are disabled when a fire is predicted in the area. For example, if the structure is predicted to be perforated in an aircraft crash, and, considering possible paths, jet fuel is predicted to penetrate various portions of the structure, then it could be conservatively assumed that all systems and components in the area are disabled and cannot function. Failure of the fire barriers, subsequent spreading of the fire to the adjacent fire areas, and the loss of SSCs within the adjacent fire areas may all be assumed as well.

This approach acts as a first level of screening, which could be refined with more information about the fire (heat, duration, smoke, etc.) and performance of the fire barrier.

5.5.2.2. Detailed analysis

A detailed analysis may be performed to remove conservatism from the evaluation process. From the fire model chosen for an internal or external fire analysis, the fire conditions (fire plume temperatures, ceiling jet temperatures, incident radiation heat fluxes) near the relevant SSCs are derived and the associated effects to the safety related items determined. Damage to the SSCs is ascertained using predetermined failure criteria.

The calculation of the temperature distribution can be based on the 'lumped temperature' concept if the component is thermally thin. This can be tested using the Biot number:

$$Bi \equiv \frac{hL}{\lambda} \tag{91}$$

where

h is the effective heat transfer coefficient;
L is the thickness of the component;

and λ is the thermal conductivity of the component.

Components with a Bi of less than 0.1 can be considered thermally thin. The temperature $T(t)$ inside such components is uniform and can be solved from the ordinary differential equation:

$$M \cdot c_p \frac{dT}{dt} = \dot{q}''(t) \tag{92}$$

where

M is the mass of the component;
c_p is the specific heat capacity;

and $\dot{q}''(t)$ is the net heat flux at the component surface, which is the primary boundary condition for thermal analysis (see Section 4.4.2).

If the Bi is greater than 0.1, the internal heat conduction needs to be resolved, taking into account the temperature dependent material properties and internal radiation within optically thin materials or air gaps. One dimensional analysis is often enough.

The cable insulation and damage thresholds depend on the type of material used. For thermoplastic cables, such as polyvinyl chloride insulated cables, it is recommended to assume a cable insulation failure temperature of 200°C [100]. For thermosetting cables, the failure temperature was found to be about 400°C. For large fires, these thresholds are not critical for fire damage calculations because the threshold levels are easily exceeded due to the intensity of the fire. The cables in fire areas are usually lost during a direct fire exposure; the resistance time is of the order of minutes rather than hours. In small fires, the threshold temperatures may significantly affect the failure time. In addition, the details of the cable function and failing circuit become important [101].

The malfunction temperatures of various electrical and electromechanical devices that are commonly used in nuclear installations have been determined [102]. For electromechanical devices, the malfunction temperatures were summarized as:

— Electrically controlled, support high temperatures: >200°C.
— Mechanically controlled alone: >170°C.
— Equipment associated to a bulb: $130 < T < 170$°C.
— Mechanically and electrically controlled: around 140°C.

For electronic boards, the lowest observed malfunction temperature was 95°C.

In addition to the thermal effect, electronic equipment is also sensitive to smoke and moisture. Smoke-induced failures are possible in smoke-filled parts of the plant where the thermal effects remain small. Smoke can cause circuit bridging and memory chip failures. Experimental studies on the performance of electronic components have shown that the combined effect of smoke and moisture is stronger than the effect of moisture or smoke alone [103]. Quantitative values for the thresholds of soot or moisture concentrations are difficult to determine. The effects of smoke can be significantly reduced by surface coating.

Appendix I

PREDICTION OF LOADING FUNCTIONS
FOR AIRCRAFT CRASHES

I.1.   GENERAL

A key parameter, which defines the effect of an aircraft missile on a target structure, is the velocity of the missile at impact. Aircraft impact speeds are a function of aircraft maximum controlled flight speed near the ground surface. For accidental aircraft impacts, the velocity normally considered is landing speed. Reference [86] gives forcing functions for a number of general aviation aircraft developed using methods similar or identical to Riera's method [7]. In the absence of sufficient data on a specific aircraft, its impact forcing function may be approximated by scaling one of these forcing functions using the procedure given in section C.6.3.3.2 of Ref. [80]. The typical impact forcing function of a military aircraft crash, resulting from a Phantom F-4F military aircraft crashing at about 800 km/h, with a missile crash load cross-sectional area equal to 7.0 m^2, can be seen in Ref. [104].

The choice of the loading (at least the aircraft type) depends on the Member State.

I.2.   MILITARY AIRCRAFT

The derivation of an impact load function for a Phantom F-4F military airplane is presented in the following, using the data provided in Table 15. The impact load function used here as an illustrative example was first derived by Drittler and Gruner [105], and was eventually used for the design of German plants [106].

The derivation of a loading function is based on available data corresponding to the aircraft. More realistic loading functions may be derived, as indicated, by detailed analyses of the impact–crash processes using refined coupled mathematical models of the impacting aircraft and the impacted structure.

The segmentation geometry of the Phantom aircraft is shown in Fig. 39, and the lengths of the different segment parts are listed in Table 16 and estimated from Fig. 39.

TABLE 15. CHARACTERISTIC DATA FOR PHANTOM F-4F MILITARY AIRCRAFT [104]

Characteristic data	Phantom F-4F
Length (L)	18 m
Impact velocity (v_o)	215 m/s
Loading area (A)	7 m²
Impact angle	90°
Mass of the fuselage (M_f)	13 000 kg
Mass of wings and fuel ($M_w + k$)	7 000 kg
Take-off weight	20 000 kg

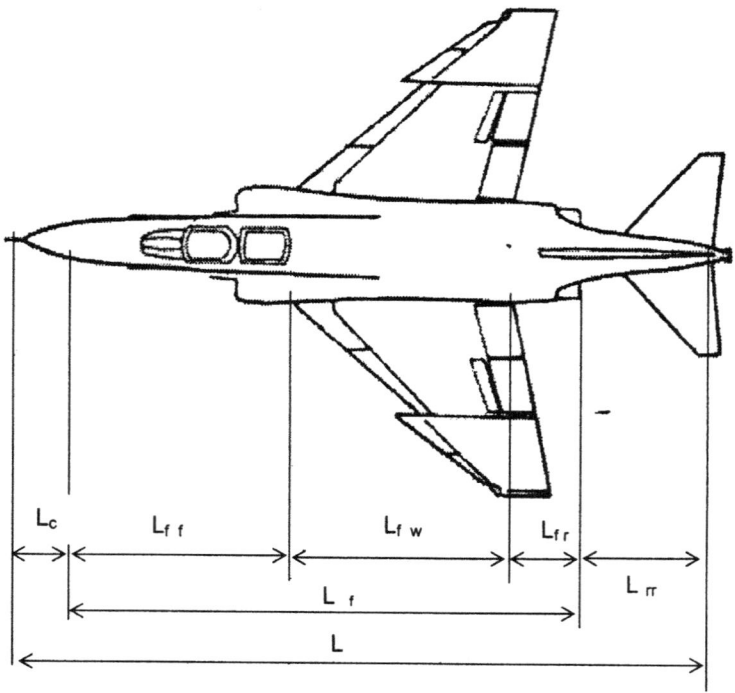

FIG. 39. Segmentation geometry of the Phantom aircraft.

TABLE 16. SEGMENTED GEOMETRY DATA OF THE PHANTOM F-4F

Segment	Partial length (m)
L_c	2
L_{ff}	6
L_{fw}	5
L_{fr}	2
L_f	13
L	18

I.2.1. Duration of impact phases

The duration of impact for each segment is calculated assuming the initial impact velocity to be 215 m/s. Further, the velocity is assumed to decrease to a value of 200 m/s. The segment crash sequence is as follows:

$$t_c = \frac{L_c}{v_c} = \frac{2}{215} = 9 \, (\text{ms})$$

$$t_{ff} = \frac{L_{ff}}{v_1} = \frac{6}{215} = 28 \, (\text{ms})$$

$$t_{fw} = \frac{L_{fw}}{v_2} = \frac{5}{200} = 25 \, (\text{ms})$$

$$t_{fr} = \frac{L_{fr}}{v_1} = \frac{2}{200} = 10 \, (\text{ms})$$

Total duration:

$$t = t_c + t_{ff} + t_{fw} + t_{fr} = 72 \, (\text{ms})$$

I.2.2. Total reaction forces

The distribution of masses for a Phantom F-4F aircraft is presented in Table 17 and the masses per unit length are computed in Table 18.

The Riera formula is used for calculating the force–time function:

$$P(t) = P_b[x(t)] + m[x(t)]v^2(t) \tag{93}$$

On the basis of experience, the crushing force P_b is usually $P_b(x) \leq 0.1P(t)$. The crushing force is assumed here to be 10% of the total force. Thus, the total force P as a function of time can be computed from the masses per unit length m and the impact velocity v as:

$$P(t) = 1.1m[x(t)]v^2(t) \tag{94}$$

Impact force values calculated for the different parts of the aircraft are listed in Table 19. The calculated loading function for the Phantom F-4F is shown in Fig. 40.

In Fig. 40, it should be noted that the force plateaus correspond to the forces in Table 19 for each section of the aircraft. The slopes connecting the plateaus depend on the length Δx along which the mass per unit length is assumed to vary from each section to the other. If the derivative of function $P(t)$ is taken:

$$\frac{dP}{dt} = 1.1 \left[\frac{dm}{dx}\frac{dx}{dt}v^2 + m2v\frac{dv}{dt} \right] \approx 1.1\frac{dm}{dx}v^3 \approx 1.1\frac{\Delta m}{\Delta x}v^3 \tag{95}$$

In this case, it has been assumed that $\Delta x = 2$ m, which gives a slope of:

$$\frac{dP}{dt} \approx 1.1\frac{\Delta m}{\Delta x}v^3 = 1.1\frac{2400-1000}{2}200^3 = 6.16 \text{ (MN/ms)} \tag{96}$$

It should also be noted that in Fig. 40, according to the assumptions made, the area under the force function (impulse) should be approximately equal to 1.1 times the initial momentum of the impacting aircraft (20 000 (kg) × 215 (m/s)).

I.2.3. Experimental validation of the loading function

In order to verify and confirm the correctness of the analytical computations and the appropriateness of the analytical procedures used to derive the loading function in Ref. [106], full scale impact tests were performed within a joint

TABLE 17. PARTICIPATION OF MASSES IN A PHANTOM F-4F AIRCRAFT FUSELAGE

Aircraft part	Phantom F-4F
Front of fuselage	$M_{ff} = 13\ 000 \times (6/13) = 6000$ kg
Wing part	$M_{fw} = 13\ 000 \times (5/13) = 5000$ kg
Rear part	$M_{fr} = 13\ 000 \times (2/13) = 2000$ kg

TABLE 18. MASSES PER UNIT LENGTH OF A PHANTOM F-4F AIRCRAFT

Aircraft part	Phantom F-4F
Front of fuselage	$m_{ff} = \dfrac{M_{ff}}{L_{ff}} = \dfrac{6000\ (\text{kg})}{6\ (\text{m})} = 1000\ (\text{kg/m})$
Wing part + wings + fuel	$m_{fw} = \dfrac{M_{fw}}{L_{fw}} + \dfrac{M_{w+k}}{L_{fw}} = \dfrac{5000\ (\text{kg})}{5\ (\text{m})} + \dfrac{7000\ (\text{kg})}{5\ (\text{m})} = 2400\ (\text{kg/m})$
Rear part	$m_{fr} = \dfrac{M_{fr}}{L_{fr}} = \dfrac{2000\ (\text{kg})}{2\ (\text{m})} = 1000\ (\text{kg/m})$

TABLE 19. IMPACT FORCE VALUES FOR DIFFERENT SECTIONS OF A PHANTOM F-4F AIRCRAFT

Aircraft part	Phantom F-4F
Front of fuselage	$P_{ff} = 1.1 \times 1000 \times 215^2 = 51\ (\text{MN})$ (55 MN in Ref. [106])
Wing part + wings + fuel	$P_{fw} = 1.1 \times 2400 \times 200^2 = 107\ (\text{MN})$ (110 MN in Ref. [106])
Rear part	$P_{fr} = 1.1 \times 1000 \times 200^2 = 44\ (\text{MN})$

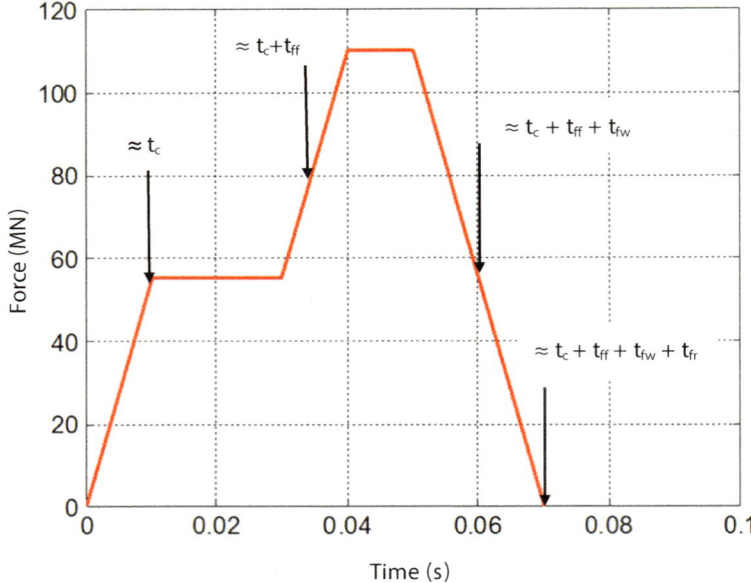

FIG. 40. Force function F(t) for the Phantom F-4F (M = 20 000 kg, v_0 = 215 m/s) (adapted from Ref. [106]).

Japanese–US research project in Albuquerque, New Mexico, United States of America in 1988 [36].

Using an existing rocket sled facility in Sandia National Laboratories, a full scale Phantom F-4D aircraft was impacted at a nominal velocity of 215 m/s into an essentially rigid block of concrete. The concrete target was 'floated' on a set of air bearings. The impact mass was 19 000 kg, which included the take-off weight, five rocket casings and 4800 kg of water that was added to simulate the fuel weight and, at the same time, provide the proper mass distribution.

The propulsion was accomplished by two stages of rockets. The aircraft impacted the target in a perfectly normal orientation and the aircraft was crushed from the front.

During impact, a short portion of each wing tip and tail was sheared off. The remainder of the aircraft, from the nose to the tail, and the engines were completely destroyed during the impact; high speed photography recorded the impact and also permitted the determination of the actual impact velocity.

The loading functions obtained during the experiment and by using the analytical method described above are shown for comparison in Fig. 41. It can be seen that, when ignoring the contribution of the rockets and the sled to the global impulse, both loading functions are very comparable.

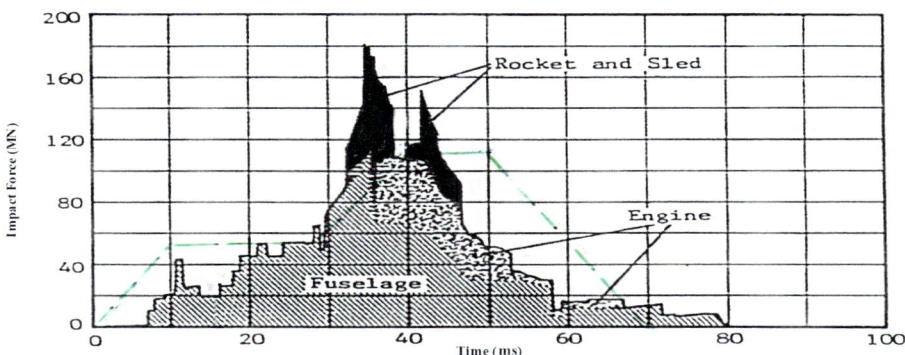

FIG. 41. Validation of the impact loading function by experiment (full scale test with Phantom F-4) (courtesy of IASMiRT [107]).

Characteristic data for other military aircraft are compiled in Table 20. It should be noted that all aircraft are supersonic. However, the reported maximum speeds cannot be attained at low altitudes. This is the reason why the impact velocities reported in Table 20 are significantly under the maximum speed of the aircraft.

TABLE 20. CHARACTERISTIC DATA FOR SOME MILITARY AIRCRAFT

Aircraft type	Dry weight of aircraft (kg)	Maximum fuel capacity (kg)	Maximum take-off weight (kg)	Maximum speed (m/s)	Impact velocity (m/s)
Starfighter F-104G	6 350	3 000	13 170	611	215
Phantom F-4F	13 700	5 100	20 000	658	215
Panavia Tornado	14 500	6 030	28 000	667	215
Eurofighter Typhoon	11 700	4 996	23 000	689	215
F/A-18C Hornet	10 400	6 370	23 500	542	215

Note: The data were compiled from several sources. Detailed information about the aircraft can be found in Ref. [108].

I.3. COMMERCIAL AIRCRAFT

The resultant loading function $F(t)$ can be calculated using the Riera formulation presented in Section 2.2 by using a simple numerical integration procedure, as follows [109]:

(1)   Set $v = v_0$, $M_r = M$, $x = 0$

where

x is the length of the crushed part of the missile;
M_r is the mass of the uncrushed part of the missile;
M is the total mass;

and v_0 is the impact velocity.

(2) $t = t + \Delta t$
(3) $x = x + v\Delta t$
(4) $F(t) = P_c(x) - m(x)v^2$ (compressive force is negative)

(5) $\Delta v = \dfrac{P_c}{M_r} \Delta t$

(6) $v = v + \Delta v$
(7) $M_r = M_r - m(x)\Delta x$
(8)   Go to (2)

The calculation ends when $x = L$ (the whole aircraft length is crushed) or $v = 0$ (the impact velocity is decreased to zero).

In the case of a commercial aircraft, the mass flow term dominates the loading function and the crushing load P_c of the fuselage part is relatively insignificant.

When assessing the mass distribution for an aircraft, the existence of the central wing tank needs to be taken into consideration. The mass of the fuel located in the central wing tank considerably affects the loading function.

Considering the design parameters of the airplane, the impact load to be expected in the event of an aircraft crash mainly depends on the mass and velocity of the aircraft. Some characteristic information on typical commercial aircraft types is compiled in Table 21. The mass of fuel is given in Section 2.4.2.2.

Bigger, heavier and faster aircraft will be developed in the future. Therefore, it may be anticipated that the load levels will increase and the crushing

TABLE 21. CHARACTERISTIC DATA FOR COMMERCIAL AIRCRAFT

Aircraft	Maximum take-off weight (kg)	Length (m)	Wing span (m)	Fuselage maximum height (m)	Fuselage maximum width (m)
Airbus A320-200	77 000	37.6	34	4.1	4.0
Boeing 767-300	156 500	55	47.6	5.3	5.0
Boeing 747-400	396 900	71	64	7.8	7.1
Airbus A380-800	560 000	72	79.8	8.4	7.14

Note: The data were compiled from the following open sources: http://www.boeing.com, http://www.chevron.com, Jane's All the World's Aircraft 2000-01, and Rolls-Royce Trent 976.

process will become more complex. Recently, considerable improvements for representing the properties of flying objects have been introduced (coupled airplane–target models) [110].

By means of such refined models of the airplanes and the impacted structure, it is possible to calculate the crash process explicitly, simultaneously considering the non-linear capabilities of the airplane and the impacted structure. However, such refined analyses are time consuming and they are reasonable only when the data for the preparation of a refined mathematical model of the airplane are reliable. In any case, based on the assumption of a rigid target condition, representative loading functions may be derived, which are useful for preliminary design purposes.

When using the Riera approach, the mass distribution m can be predicted in different ways:

— The mass distribution is scaled according to the mass distribution of a Boeing 707 presented in Ref. [109].
— The mass distribution calculated in (a) is used and, additionally, the wings are assumed to break at the outside of inboard engines, according to Ref. [7].
— A more detailed mass distribution is predicted according to the available information. This may be necessary, especially in the case of a central wing tank.

The crushing force P_c can also be assessed using different approximations:

— The crushing load distribution is scaled according to the crushing load distribution of a Boeing 707 obtained from Ref. [109].
— The same assumption is used for the crushing force (10% of the total force) as presented for the military aircraft above. It should be mentioned that this assumption is somewhat overly conservative.
— The crushing force is predicted using separate models describing the structural stiffness and strength of the aircraft.

As an example, loading functions for a Boeing 747-400 using these different kinds of assumption are predicted below. The main dimensions of a Boeing 747-400 commercial aircraft are shown in Fig. 42 and the positions of the fuel tanks are shown in Fig. 43.

The maximum take-off weight (MTOW) of a Boeing 747-400 is 396 893 kg. The maximum amount of fuel is 175 652 kg. The locations of the fuel tanks are shown in Fig. 44. After a flight time of 1.5 h, the fuel consumption is about 18 249 kg, and the remaining mass of the aircraft is 378 644 kg. The amount of fuel in various tanks at that moment is:

— Centre wing tank: 46 840 kg (89%).
— Main 2 and main 3: $2 \times 33\ 853 = 67\ 706$ kg (innermost wing tanks) (88%).
— Main 1 and main 4: $2 \times 12\ 231 = 24\ 462$ kg (89%).
— Reserve 2 and reserve 3: $2 \times 4054 = 8108$ kg (100%).
— Stabilizer: 9916 kg (at the rear of the aircraft).
— Miscellaneous: 371 kg.

This gives a total mass of fuel of 157 403 kg. The mass of fuel in the wings after a flight of 1.5 h is $2(4054 + 12\ 231 + 33\ 853) = 100\ 276$ kg.

The length of the centre wing tank is assumed to be 5.2 m, giving a mass per unit length of 46 840 (kg)/5.2 (m) = 9008 (kg/m).

The mass of the landing gear LG is about $0.044\text{MTOW} = 17\ 463$ kg. The mass of the nose assembly is $0.11\text{LG} = 1921$ kg and the mass of the main assembly is $0.89\text{LG} = 15\ 542$ kg. The structural mass of the main landing gear is $0.5\text{LG} = 8732$ kg, while the structural mass of the nose landing gear is $0.07\text{LG} = 1222$ kg.

The wing landing gear of the main landing gear is situated sideways in the flight position, while the body landing gear is in the lengthwise direction with the wheels (four in each unit) pointing forward. The nose landing gear is in the flight direction with the two wheels pointing forward.

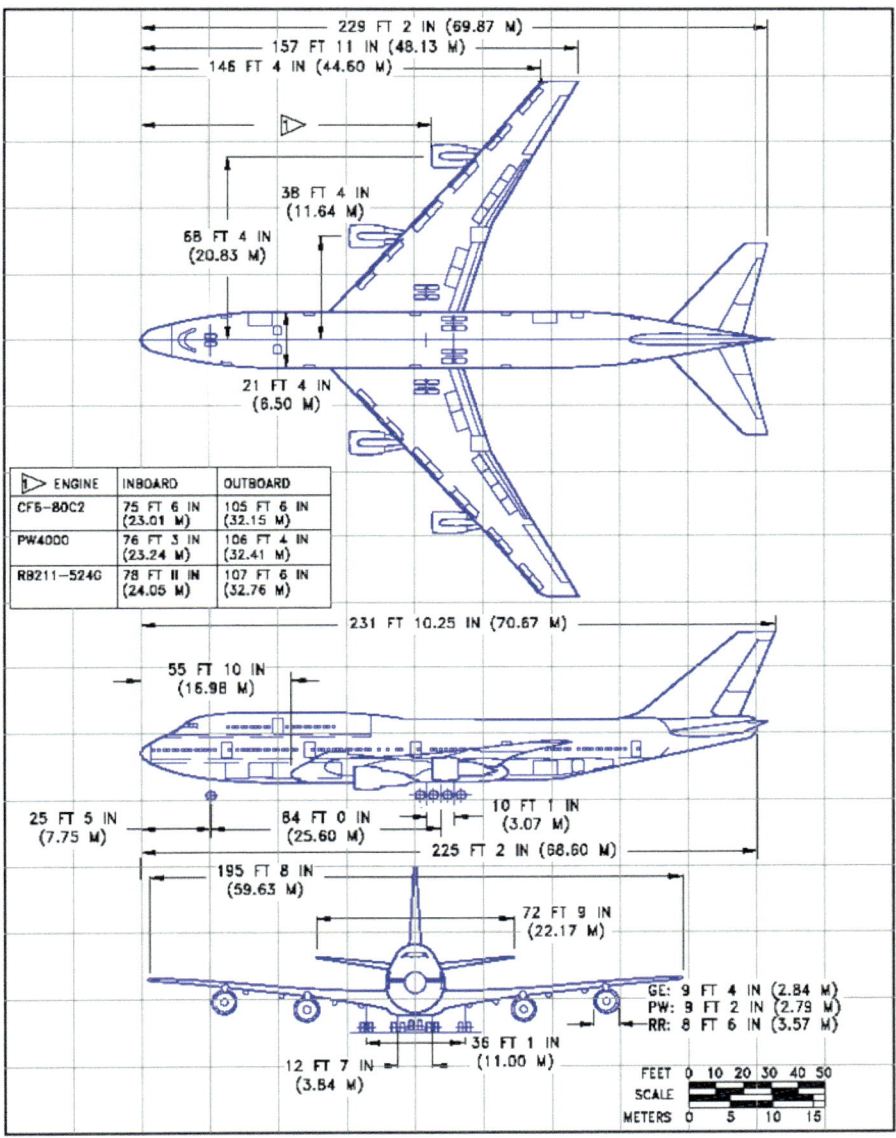

FIG. 42. Boeing 747 commercial aircraft (© 2002 The Boeing Company. All rights reserved).

The mass of the main landing gear is assumed to be distributed over a distance of 3 m, and the mass of the nose landing gear, over an interval of 1.5 m.

The mass of an engine, e.g. GE CF6, is about 4300 kg, and the length of this engine is 4.267 m. The mass of an engine, including some additional gear,

141

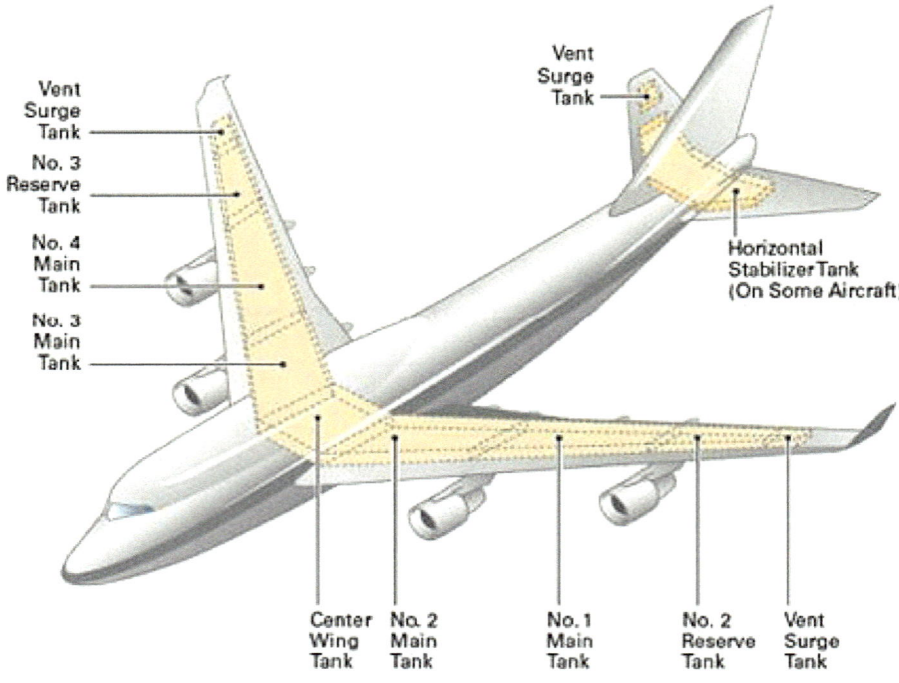

FIG. 43. Fuel tanks in a Boeing 747-400 (www.energy-visions.com).

is 5915 kg and the mass of the nacelle is 1298 kg. Adding these masses together gives 7213 kg, so that the mass of one pair of engines is about 14 426 kg (engine and nacelle).

The structural mass of the wings, including the engines, is 79 028 kg. Subtracting the masses of fuel, wings (including engines) and landing gear from the MTOW: 396 893 – 175 654 – 79 027 – 17 463, yields an estimate of 124 749 kg for the mass of the fuselage.

Based on these data, a mass distribution for the Boeing 747-400 is determined and shown in Fig. 44 as case sl (red line).

Case slr in Fig. 44 (blue discontinuous line) assumes that the wings break at the outside of the inboard engines and that the torn-off mass no longer influences the motion when the crushing length becomes 30.3 m. In this case, the lost mass is composed of the following parts:

— Fuel in main 1 and main 4: 24 462 kg.
— Fuel in reserve 2 and reserve 3: 8108 kg.
— Wing outside inboard engines: 17 182 kg.

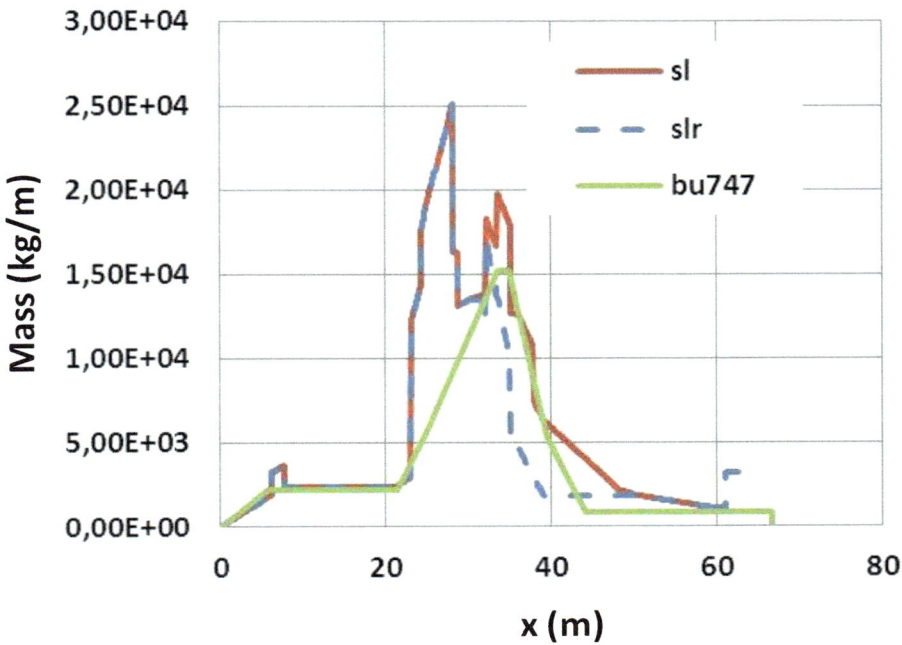

FIG. 44. Assumed mass distributions for a Boeing 747; x is the distance from the nose.

— Two engines (nacelle and propulsion): 2596 + 11 830 = 14 426 kg (for one engine: 7213 kg).

This gives a total of 64 178 kg for the break-off mass. Here it is assumed that the fuel in the innermost wing tanks is not included in the torn-off mass. The break-off point is assumed to be 30 m from the nose.

The mass distribution obtained by scaling the total mass of the aircraft according to the mass distribution of the Boeing 707 from Ref. [7] is also shown in Fig. 44 (case bu747 (green line)). By this scaling method, the tear-off mass becomes larger than that calculated by the method described above.

The corresponding crushing load distributions P_c are presented in Fig. 45, for cases sl and slr.

Figure 46 shows the impact force computed for a Boeing 747-400 travelling at a speed of 100 m/s and impacting a rigid target, sl_100 (sl stands for step wise linear mass distribution). Curve slr_100 is calculated by assuming that the wings break at the outside of the inboard engines, as assumed also in Ref. [7]. For comparison, the estimate from a calculation model based on Ref. [7], bu747, is also given.

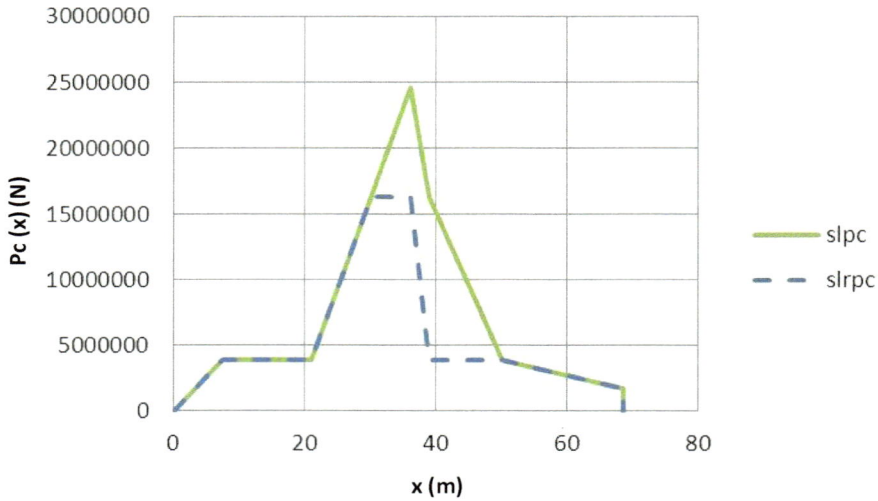

FIG. 45. Assumed crushing force distribution for a Boeing 747; slrpc excluding tear-off mass, and slpc; x is the distance from the nose.

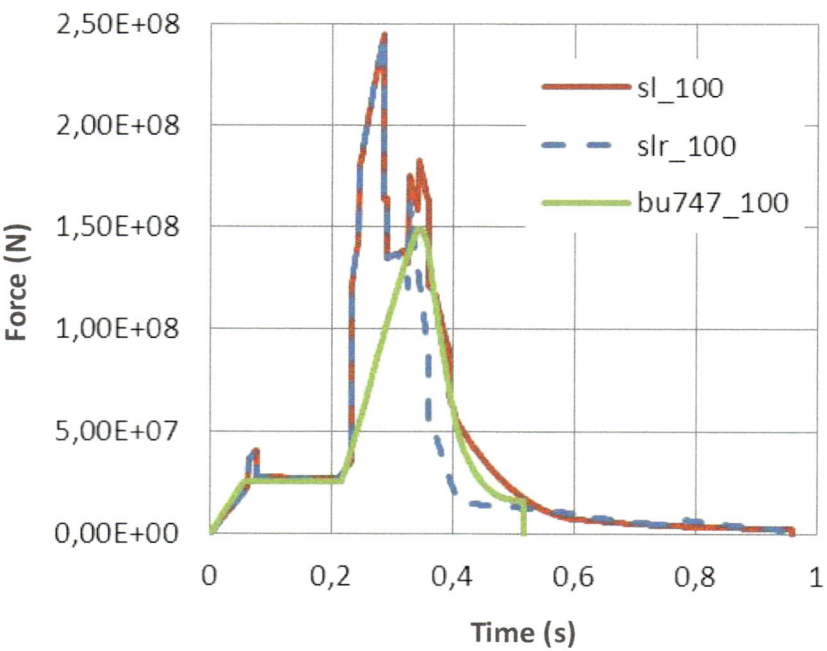

FIG. 46. Force resultants F(t) for a Boeing 747-400; v_0 = 100 m/s.

The predicted loading function is to be applied to the structure under consideration. The resultant force will be distributed on an approximate loading area. The loading function needs to be divided into at least two separate force–time functions (fuselage and wing part) and these are to be applied to corresponding impact areas. Owing to crushing and deformation of the airplane cross-section, the actual impact load area is somewhat larger than the actual cross-section of the aircraft. An example of the assessment of the impact loading area is shown in Fig. 47.

The cross-section area of the fuselage is A = 40 m² and the effective radius is r_{eff} = 3.56 m. The effective radius of the crushed cross-section is assumed to increase by 15% and, thus, r_{mod} = 1.15r_{eff} and the actual impact area is A_{mod} = 52 m². The loading area can be modified if properly justified.

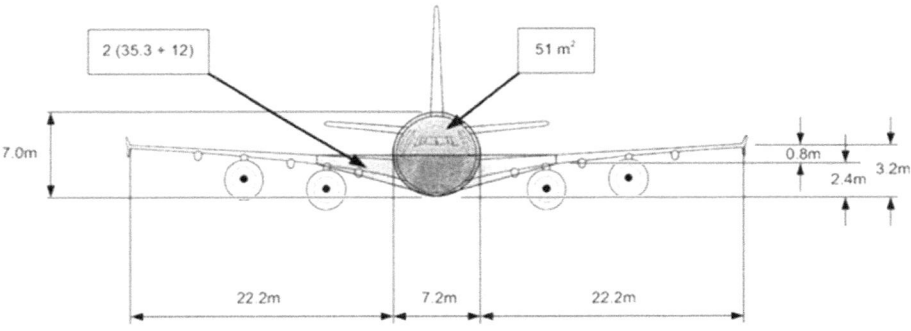

FIG. 47. Loading areas for a Boeing 747.

I.4. CHARACTERISTIC DATA FOR ENGINES AND LANDING GEAR

Engines and landing gear cause semi-hard missile impacts during an aircraft crash. The shaft of an engine can be idealized as a rather thick walled tube. This type of a compact structure can be considered as a hard missile. Some characteristic data on engines are compiled in Table 22.

The total mass of the main landing gear and nose landing gear assemblies is roughly 4.4% of MTOW. The wing landing gear of the main landing gear is situated sideways in the flight position, while the body landing gear is in the lengthwise direction, the wheels (four in each unit) pointing forward. The nose landing gear is in the flight direction, the two wheels pointing forward. During

TABLE 22. CHARACTERISTIC DATA FOR ENGINES

Engine type	Dry mass (kg)	Fan diameter (m)	Engine length (m)	Aircraft (number of engines)
CFM56-5B4	2380	2.4	2.6	Airbus A320 (2)
GE CF6-80C2	4300	2.7	4.3	Boeing 767 (2), Boeing 747 (4)
Rolls-Royce Trent 900	5800	2.9	4.3	Airbus A380 (4)
GE-J79	1764	0.76	3.1	Phantom F-4 (2)

the flight, the landing gears are packed in a rather compact form and since they are mainly steel structures, they create at least a considerable semi-hard missile during an aircraft crash. The effect can clearly be seen in fig. 7.9 of Ref. [111]. Some typical landing gear layouts are shown in Fig. 48.

FIG. 48. Typical landing gear layouts [113].

The landing gear mass can be estimated from aircraft take-off weight by statistical weight equations [112]. Mass estimates for landing gears of some typical aircraft are given in Table 23.

TABLE 23. CHARACTERISTIC DATA FOR LANDING GEARS

Aircraft type	Maximum take-off weight (kg)	LG = 0.044MTOW	LG_{nose} = 0.11LG	$Gear_{nose}$ = 0.07LG	LG_{main} = 0.89LG	$Gear_{main}$ = 0.5LG
Airbus A320	77 000	3 388	373	237	3 015	1 694
Boeing 767	156 500	6 886	757	482	6 129	3 443
Boeing 747	396 900	17 463	1 921	1 222	15 542	8 732
Airbus A380	560 000	24 640	2 710	1 725	21 930	12 320

Note: LG: landing gear; MTOW: maximum take-off weight.

Appendix II

SIMPLIFIED METHODS FOR STRUCTURAL IMPACT

II.1. EXAMPLE OF APPLICATION OF A TWO DEGREE OF FREEDOM MODEL

An illustrative example is shown in Fig. 49. It depicts the roof of a reinforced concrete protection cover of a nuclear processing plant with thickness varying from 1.8 to 2.0 m and a main span of 26 m. The span in the transverse direction is considered to be 'large'. The slab is clamped on all four sides. It is hit by an aircraft in the middle of the roof.

The TDOF model, as described in Section 4.2.2, is used for the basic design of the structure. The masses are calculated according to Eqs (54, 55). The total effective mass m_e of the slab is derived by comparison of the first frequency of the two-mass system with the bending dominated frequency of the slab. The mass of the punching cone is calculated with Eq. (54) with an angle $\alpha = 45°$.

The non-linear characteristics of the springs are shown in Fig. 50 for the bending spring r_1 and for the shear spring r_2. Both springs are modelled by trilinear load–deformation curves. The first part represents the initial linear behaviour until cracking of the concrete. The second part accounts for the reduced stiffness of the cracked concrete and the last part, with ideal plastic behaviour, stands for yielding of the reinforcement. The limit load of 85 MN of the bending spring is calculated according to the yield line theory. The limit load of 100 MN of the shear spring is calculated with Eq. (53) for the stirrups. Failure is defined by limit deformations under bending or shear. The deformation under bending is limited by the allowable rotations in the yield line field. The deformation under shear is limited by the ultimate strain of the stirrups.

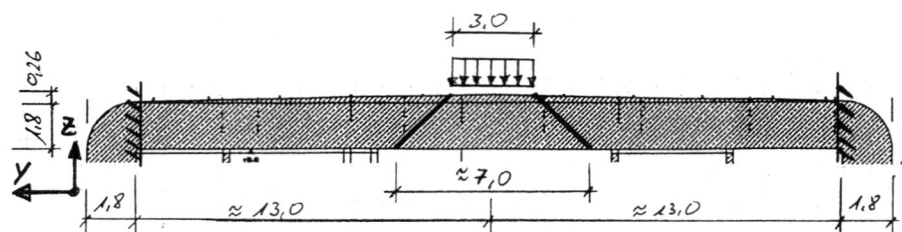

FIG. 49. Illustrative example: airplane crash on a roof (courtesy of IASMiRT [114]).

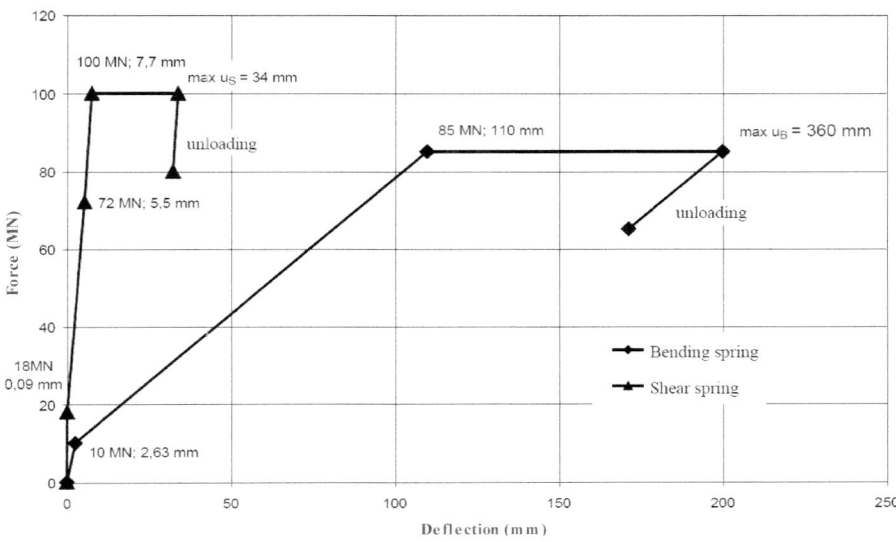

FIG. 50. Non-linear characteristics of the springs (courtesy of IASMiRT [114]).

The model is loaded by the time history of a military (Phantom) airplane crash. The response of the system is calculated by a non-linear time history analysis. The required result is the time history of the deformation of the system (Fig. 51). The green curve corresponds to the deformation of mass 1 (bending mass), and the red curve to the deformation of mass 2 (shear mass). The blue curve is the difference of the deformations of mass 1 and mass 2. This curve represents the deformation of the shear spring. The strength of the slab is sufficient if the maximum deformation of the bending spring (green curve) as well as the maximum deformation of the shear spring (blue curve) is below the limit deformations. This condition is met here. It is remarkable that the maximum deformation occurs in the reverberation time after the end of the airplane crash load. The load history ends at 0.07 s.

As a by-product of the analysis, the spring forces are also calculated (Fig. 52). The green curve in Fig. 52 is the force in the bending spring and the red curve is the force in the shear spring. The history of the shear spring, which is comparably stiff, basically follows the load history with some oscillation after the end of the load. The history of the bending spring is characterized by the constant force at the plastic limit force of 85 MN, with a decrease after the maximum deformation is reached. It is noted that the unloading phase in the selected characteristics of the springs is only roughly approximated. The main aim of the analysis was the calculation of the maximum deformations. The reverberating

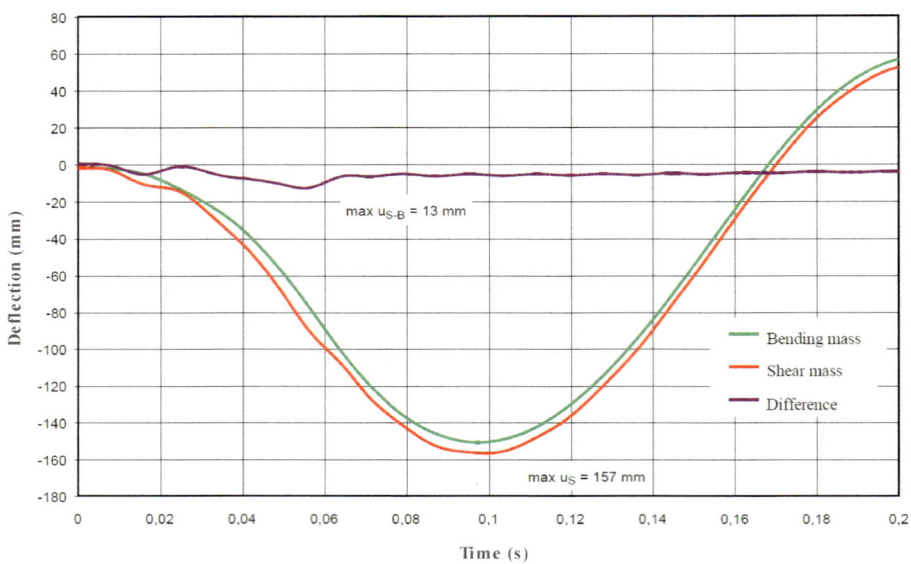

FIG. 51. *Deformations of the two degree of freedom system (courtesy of IASMiRT [114]).*

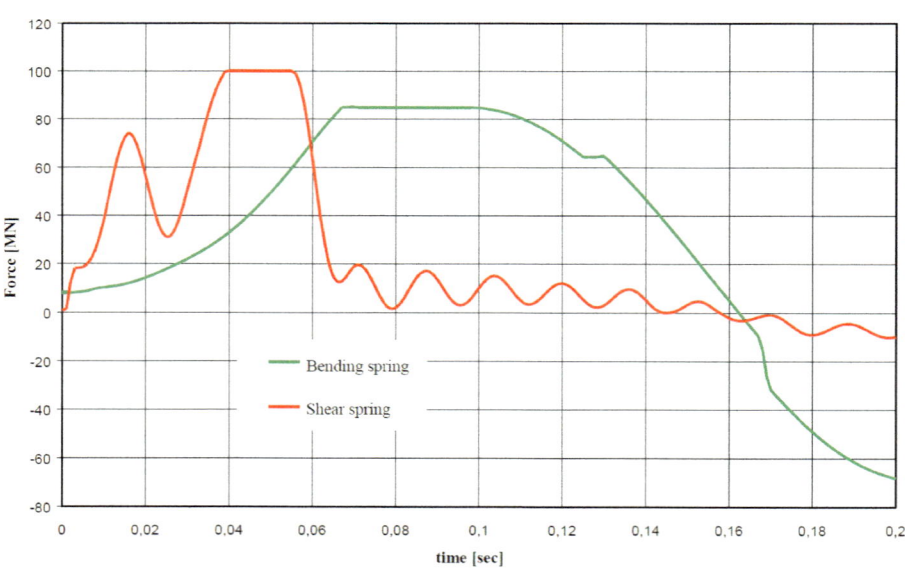

FIG. 52. *Spring forces in the two degree of freedom system (courtesy of IASMiRT [114]).*

phase of the vibrations was not in the scope of this analysis. Nevertheless, the backward vibration with reversal of the force can be observed within the analysed time range of 0.2 s. Owing to this backward vibration, sufficient reinforcement at the front side of the target is required. Most codes demand the same amount of reinforcement on both sides of the target.

The TDOF model is generally sufficient for the basic design of plane concrete structures with simple geometry and boundary conditions, before starting with detailed finite element analysis. Additionally, it is useful to check, for an existing plant, whether the available reinforcement is sufficient to sustain an aircraft crash [115].

II.2. DERIVATION OF VERIFIED LOADING FUNCTIONS

The overall dynamic response of a structure impacted by missiles or an aircraft could be easily derived using linear-elastic finite element models of the structure and performing analyses in the time domain.

However, in the impact zone, the outer shells or walls can experience a strongly non-linear local behaviour. Owing to this fact, the loads finally acting on the rest of the structure can change significantly. To take these effects into consideration, detailed non-linear analyses need to be carried out using refined finite element models capable of sufficiently describing the linear and non-linear behaviour of the reinforced concrete structures (Fig. 53).

The main results of a non-linear analysis of the impacted structure are time histories of displacements, forces and moments obtained for the characteristic cross-sections of the impacted target. On the basis of the internal forces and moments obtained, the dynamic response of the whole excited structure could subsequently be derived introducing the previously obtained time histories of forces and moments in the significant intersection of the linear-elastic global model.

However, difficulties are often present when introducing the input motion defined by three forces and two moments into the corresponding significant intersection. Thus, a procedure for derivation of VLF (verified for the local capabilities of the structure, such as geometry, wall thickness, reinforcement, and material properties) was established in order to cover the non-linear effects.

The goal is to generate a load function which, if acting on the same finite element local model possessing linear-elastic capabilities (Fig. 53(b)), results in time histories which are in good agreement with the time histories obtained on the basis of non-linear analysis and model assumptions (Fig. 53(a)), and which also take care of the balance of energy.

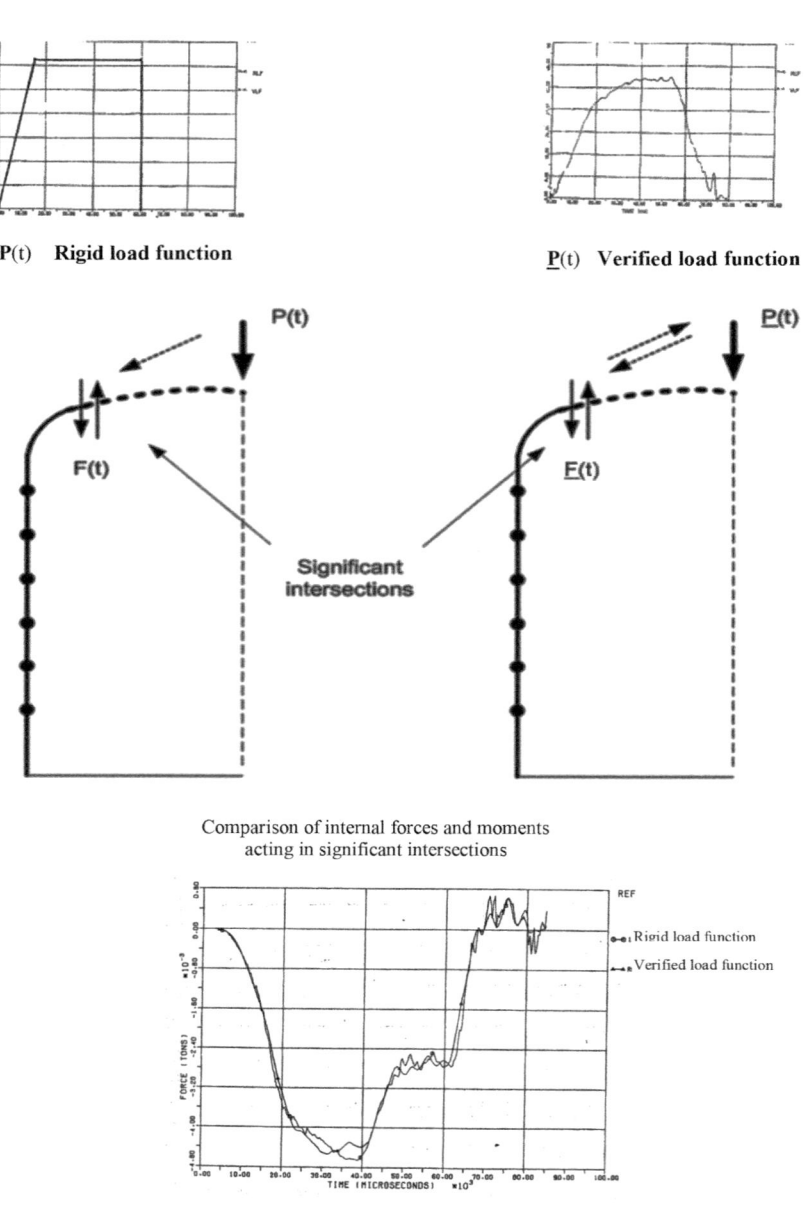

P(t)   **Rigid load function**

P̲(t)   **Verified load function**

Significant
intersections

Comparison of internal forces and moments
acting in significant intersections

o—o₁ Rigid load function

▲—▲₂ Verified load function

(a) Non-linear analyses

(b) Linear-elastic analyses

FIG. 53. Non-linear and linear finite element models of a reactor building (forces acting in cross-section).

The derivation procedure of the verified load is presented below [53]. The procedure for derivation of VLFs was approved and thoroughly benchmarked by the authorities. within many licensing processes of the design of nuclear installations in Germany.

II.2.1. Theoretical approach

An arbitrary general loading history $p(t)$, specifically the intensity of loading $p(\tau)$ acting at time $t = \tau$, is presented in Fig. 54. This loading acting during the short interval of time $d\tau$ produces a short duration impulse $p(\tau)d\tau$ and the response of the structure to this impulse can be evaluated in terms of displacements, stresses, forces, etc. The entire loading history may be considered to consist of a succession of such short impulses, each producing its own response.

For linear-elastic systems, the total response F can be obtained by summing all of the responses developed by the impulses during the loading history. When the duration of the impulses approaches zero, the exact response can be written in the form of a Duhamel integral or convolution integral:

$$F(t) = \int_0^t p(\tau) f(t - \tau) dt \tag{97}$$

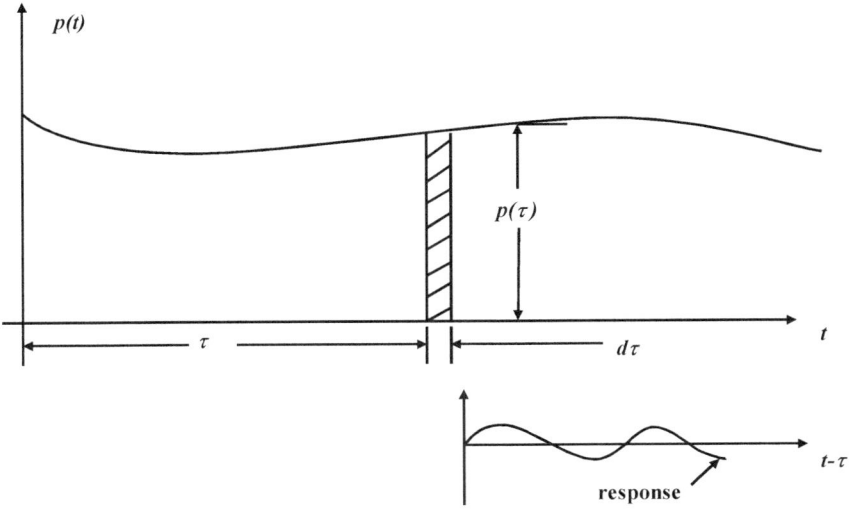

FIG. 54. Arbitrary general loading history.

The function $f(t - \tau)$ is referenced as the unit impulse response and it expresses the response of the linear-elastic system (damped or undamped) to a unit impulse applied at time $t = \tau$.

II.2.2. Method of resolution

Equation (97) can be used to obtain the loading function $p(t)$ which, acting on the linear-elastic model, would produce the internal forces and moments computed for the intersection using the non-linear analysis. This is the VLF.

In order to solve Eq. (97), the loading $p(t)$ can be approximated by a succession of triangular loadings of finite duration $2\Delta t$ and magnitude $p(\tau)$ applied at time $t = \tau$ and giving a response of $f(t - \tau)$ at the specified location. For convenience in the numerical calculation, the functions f and F are evaluated using the same time increment Δt.

The input time histories of internal forces and moments derived from the non-linear analysis can be approximated by a succession of triangular loadings of specified duration at the specified location. Two or more input time histories can be run at the same time.

At time $t = N\Delta t$, Eq. (97) can be approximated by:

$$F_N = \sum_{i=0}^{N-1} p_i f_{N-i} \tag{98}$$

The intensity p of the loading function that is looked for is obtained from Eq. (98) by the following recurrence relation:

At time $t = N\Delta t$:

$$p_N = \frac{1}{f_1} \left[F_{N+1} - \sum_{i=0}^{N-1} p_i f_{N+1-i} \right] \tag{99}$$

with $p_0 = F_1/f_1$.

In practice, since there are several internal forces and moments at the significant intersection, Eq. (99) is solved using a least squares method, which minimizes the difference at each time for all the relevant internal forces and moments, and reduces the oscillations between two consecutive values of p_i:

$$D_N = F_N - \sum_{i=0}^{N-1} p_i f_{N-i} \tag{100}$$

II.2.3. Procedure for determination of equivalent loading

The following four steps are used to determine the equivalent loading function p for the global linear model:

(a)   Specification of the function $F(t)$: The function F is the time history of the non-linear internal forces in the significant intersection of the local concrete model.

(b)   Evaluation of the unit function $f(t)$: The response to a unit triangular loading of duration $2\Delta t$ is computed in terms of internal forces at the significant intersection. The structure is represented between the location of the impact and the mentioned intersection by refined elements and by less refined elements in the rest of the structure. The selected duration of the impulse Δt is made compatible with the maximum frequency of interest (here assumed to be 80 Hz).

(c)   Determination of the loading function $p(t)$: Knowing $F(t)$ and $f(t)$, the equivalent loading is computed using Eq. (99). For this purpose, a post-processor needs to be developed following the method described in Section II.2.2.

(d)   Verification of the solution: The computed loading is applied to the finite element model described in (b) and the response is compared to the function $F(t)$.

II.2.4. Examples of verified loading function application

The presented calculation method provides a way to compute an equivalent excitation which, applied to a global linear-elastic model, will allow for computation of the global structural response (Fig. 55).

As discussed, the procedure of developing modified load functions is given in two steps:

(a)   Non-linear dynamic calculations applying the RLF and using refined finite element models for the impacted area to derive the internal forces at significant sections, adjacent to the non-linear zone in which the reinforcing steel remains linear-elastic (Figs 56 and 57);

(b)   Computation of a VLF, which, applied to a linear-elastic model of the structure, induces internal forces in the significant section which are in good agreement with those forces and moments determined for this region by means of non-linear analyses obtained in (a) (Fig. 58).

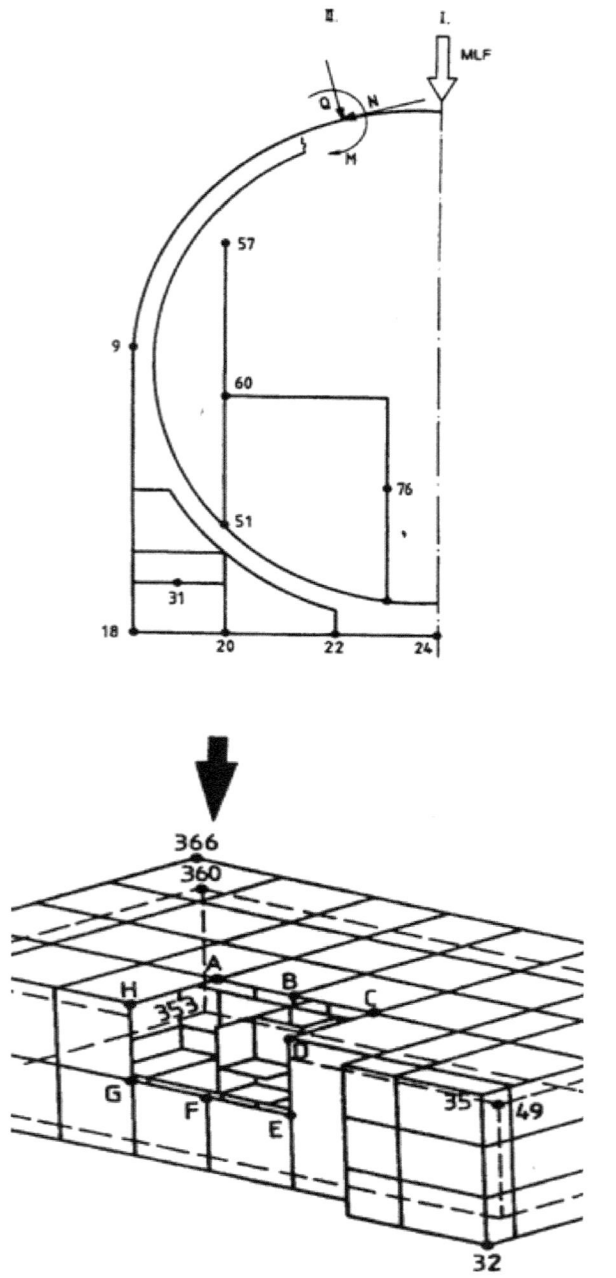

FIG. 55. Excitation of (a) an axi-symmetric and (b) a box-shaped structure using time histories of internal forces and moments in the significant intersection.

156

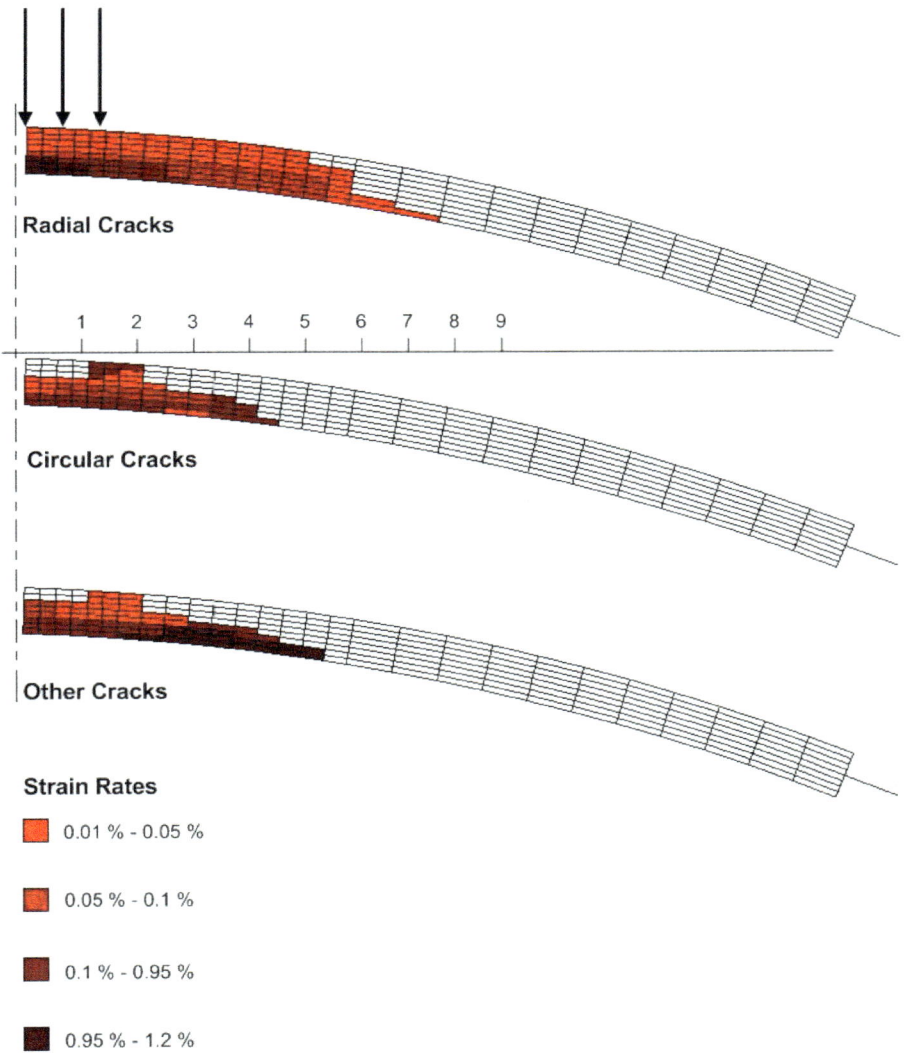

Radial Cracks

1 2 3 4 5 6 7 8 9

Circular Cracks

Other Cracks

Strain Rates

■ 0.01 % - 0.05 %

■ 0.05 % - 0.1 %

■ 0.1 % - 0.95 %

■ 0.95 % - 1.2 %

FIG. 56. Crack distribution and significant intersection on the dome of a pressurized water reactor containment building.

Based on the time histories of forces and moments obtained by non-linear calculations, VLFs were determined for typical impact regions of a pressurized water reactor containment building. Representative examples are given in Figs 59–61. These figures show that the VLF will generally have, in relation to the specified RLF, a longer duration and lower load level.

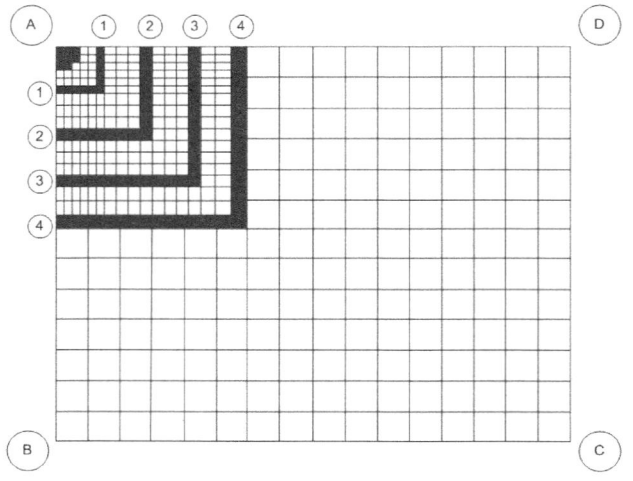

FIG. 57. Significant intersections on wall or cylindrical shell models.

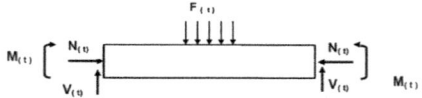

The verified load function is an updated load function which under linear-elastic status of stresses in the impacted structure provides in the considered intersection the same internal forces N(t), V(t) M(t) as the rigid loading function under realistic nonlinear behaviour of the target.

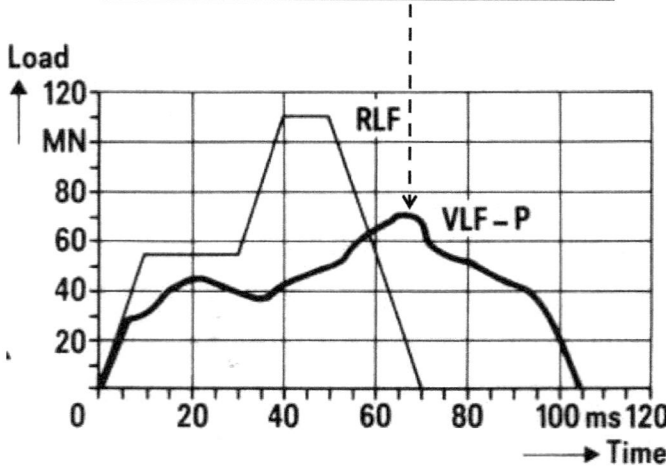

FIG. 58. Balance of forces in the significant interface of a plate. $F(L) \gg F(N)$; $N(L) \gg N(N)$; $V(L) \gg V(N)$; $M(L) \gg M(N)$.

158

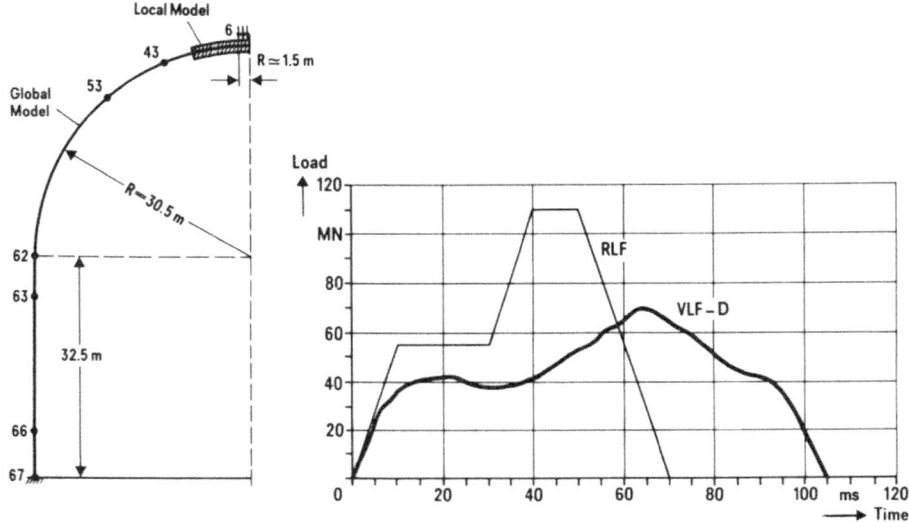

FIG. 59. *Verified loading function (VLF-D) for a crash on the dome region (courtesy of IASMiRT [53]).*

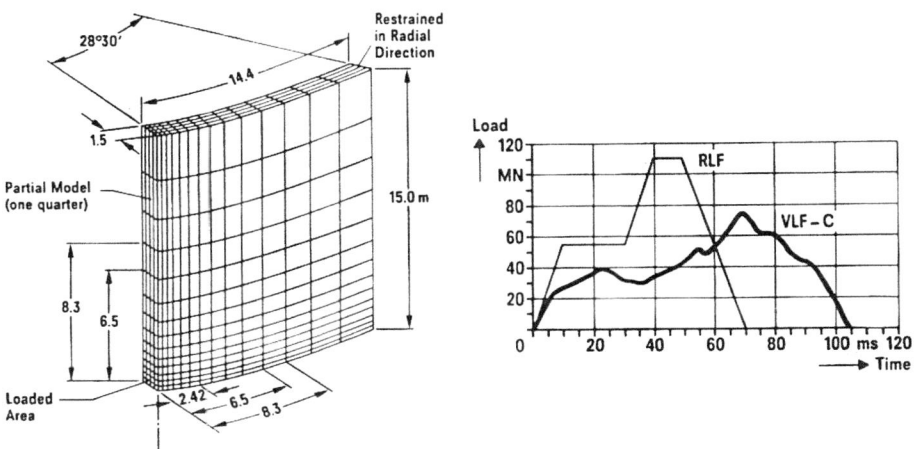

FIG. 60. *Verified loading function (VLF-C) for a crash on the cylindrical shell (courtesy of IASMiRT [53]).*

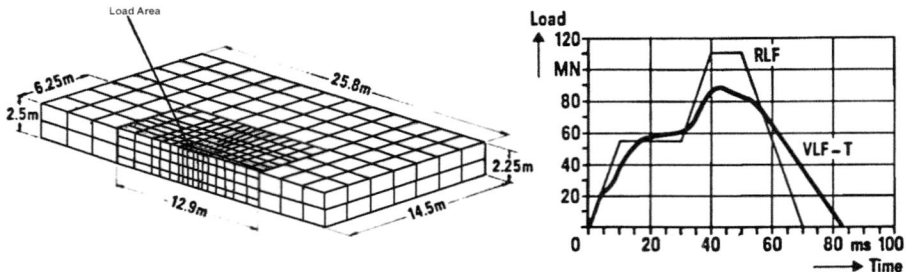

FIG. 61. Verified loading function (VLF-T) for a crash on a wall/roof region (courtesy of IASMiRT [53]).

II.3. TYPICAL DYNAMIC RESPONSE SPECTRA

The load function for the military aircraft derived in Appendix I has to be applied according to the definition as acting on a (circular or rectangular) surface of 7 m² of the building model. According to the various impact locations shown in Figs 62 and 63, derived in preliminary studies, a corresponding number of dynamic responses need to be calculated for supporting points of safety relevant components and systems.

Examples of the results of calculations performed according to the assumptions and mathematical models described above, using RLFs and VLFs, respectively, are shown in Figs 64 and 65.

A comparison of examples of floor response spectra for the axi-symmetric building (Fig. 64) shows that the spectra determined on the basis of VLFs are characterized by a 30% to 60% reduction of acceleration values and a shift in the frequency range.

Comparing the acceleration spectra obtained for a rigid box-shaped building (Fig. 65), it can be recognized that the acceleration spectra determined on the basis of the VLFs (VLF-EC) are about 10%–15% lower in the low frequency range and up to approximately 30% lower in the high frequency range than the corresponding values obtained by the RLF.

160

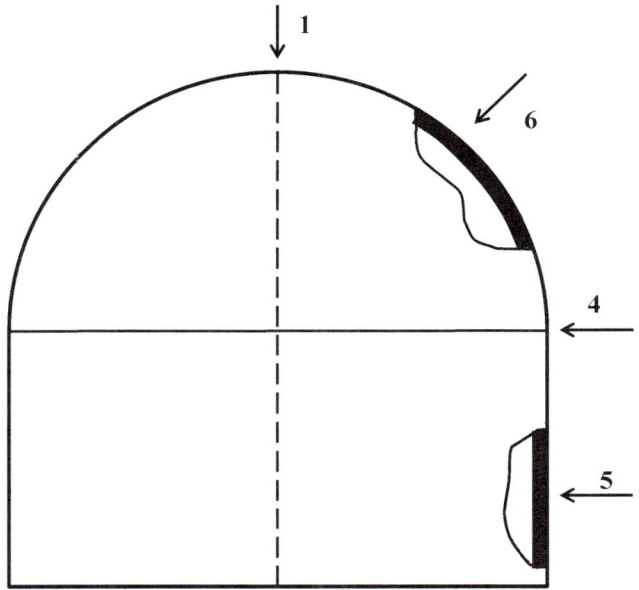

FIG. 62. Characteristic impact regions on an axi-symmetric building (courtesy of IASMiRT [50]).

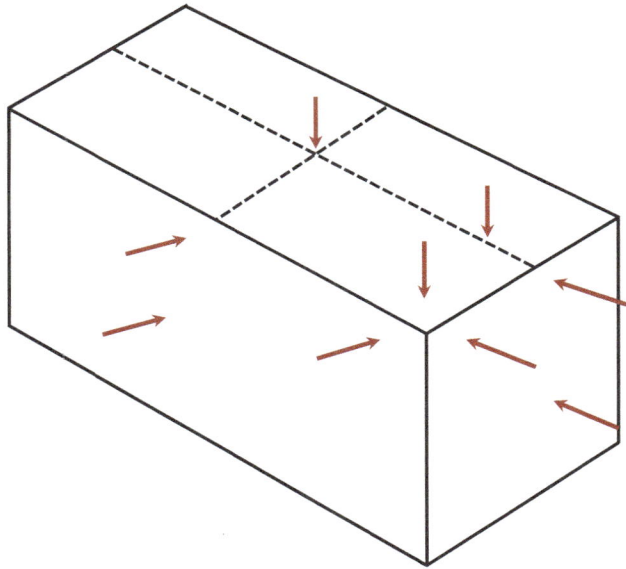

FIG. 63. Characteristic impact regions on a rectangular building (courtesy of IASMiRT [50]).

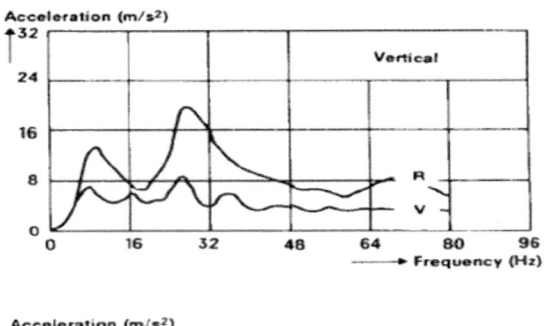

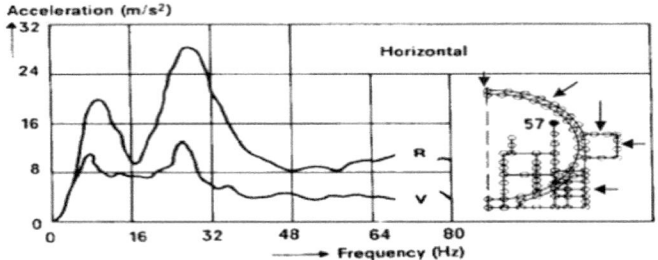

FIG. 64. *Dynamic response on the inner structure of a typical reactor building. R: rigid loading function; V: verified loading function (courtesy of IASMiRT [53]).*

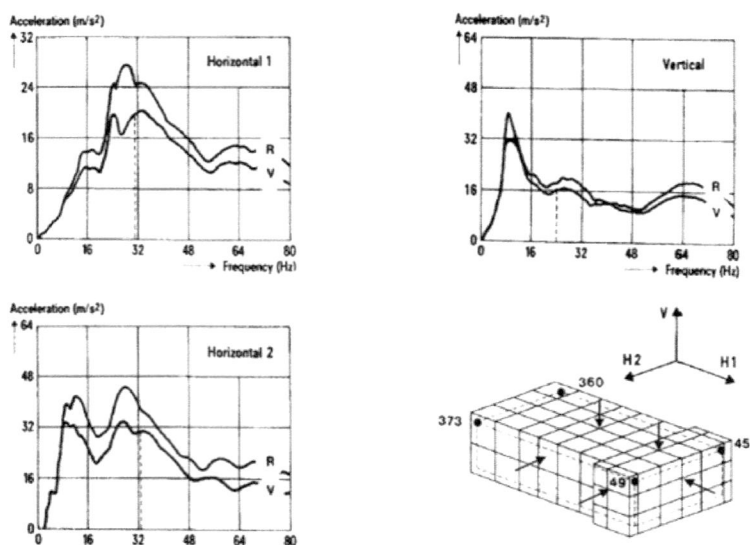

FIG. 65. *Dynamic response on the inner structure of the rectangular building. R: rigid loading function; V: verified loading function (courtesy of IASMiRT [53]).*

Appendix III

GUIDELINES FOR DESIGN AND ASSESSMENT OF CONCRETE ELEMENTS AGAINST EXPLOSION

This appendix provides numerical examples of the simplified method presented in Section 4.3.3 that can be applied in designing concrete structural elements that will withstand explosion loading. After discussing impulsive and quasi-static/dynamic structural responses, two design examples accounting for blast loading are provided. More detailed information for analysis and design of structures resistant to blast loading can be found, for example, in Refs [9, 44, 116].

III.1. RESPONSE TO IMPULSE LOADING

Figure 66 shows a triangular force–time function with a zero rise time acting on an SDOF system with mass M. The load is considered to be 'impulsive' when the duration of the applied load t_d is short compared to the response time t_m of the system (i.e. the time for the element to reach the maximum transient deflection X_m). A rule of thumb is that the load can be considered impulsive when $t_m \geq 3t_d$.

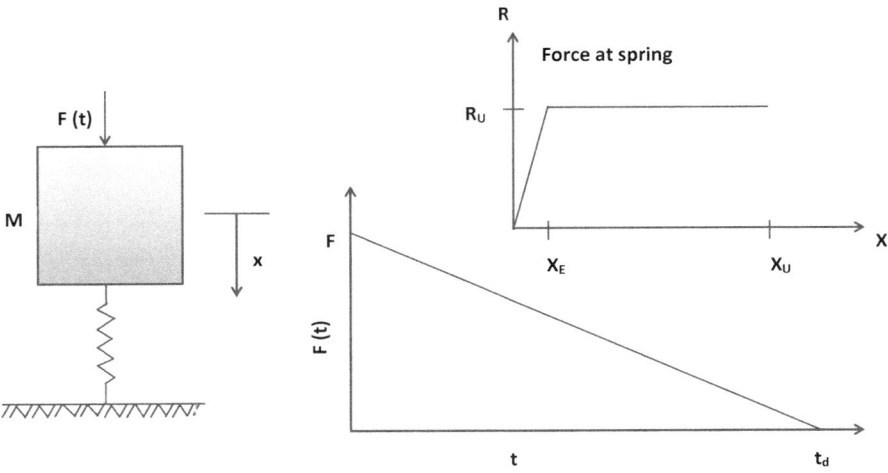

FIG. 66. Single degree of freedom elastic-plastic structure (adapted from Ref. [117]).

If F is the peak force, the impulse corresponding to the triangular pulse is:

$$I = \frac{1}{2} F \cdot t_d \tag{101}$$

The maximum deflection of the system X_m can be obtained by equating the kinetic energy delivered by the impulsive load to the strain energy developed by the member in defecting to X_m:

$$\frac{I^2}{2M} = \frac{R_u X_E}{2} + R_u (X_E - X_M) \tag{102}$$

A blast load can usually be idealized as a triangular pressure–time function with zero rise time. The parameters F and M for the SDOF idealization, displacement X and resistance R are obtained according to the particular component and failure mode being studied, as shown in the following sections.

In the SDOF idealization, deflection X is usually taken as the maximum deflection of the structural element, which is commonly modelled as a beam or frame ('one way' element) or as a plate ('two way' element). On the other hand, as explained in Section 4.3.3.2, the total real mass of the element M_R has to be multiplied by a mass factor K_M to obtain the lumped mass M of the equivalent SDOF system:

$$M = K_M M_R \tag{103}$$

Similarly, the total actual load on the element F_R has to be multiplied by a load factor K_L to obtain the concentrated load F on the equivalent SDOF system:

$$F = K_L F_R \tag{104}$$

The same applies to the resistance:

$$R_u = K_L R_{Ru} \tag{105}$$

where R_{Ru} is the actual ultimate resistance of the element, expressed in the same terms as F_R.

Factors K_L and K_M can be obtained from Table 24 for one way elements with different load and boundary conditions. Introducing Eqs (103–105) into Eq. (102) results in:

TABLE 24. TRANSFORMATION FACTORS FOR ONE WAY ELEMENTS [9]

Edge conditions and loading diagrams	Range of behaviour	Load factor K_L	Mass factor K_M	Load mass factor K_{LM}
	Elastic	0.64	0.50	0.78
	Plastic	0.50	0.33	0.66
	Elastic	1.0	0.49	0.49
	Plastic	1.0	0.33	0.33
	Elastic	0.58	0.45	0.78
	Elasto-plastic	0.64	0.50	0.78
	Plastic	0.50	0.33	0.66
	Elastic	1.0	0.43	0.43
	Elasto-plastic	1.0	0.49	0.49
	Plastic	1.0	0.33	0.33
	Elastic	0.53	0.41	0.77
	Elasto-plastic	0.64	0.50	0.78
	Plastic	0.50	0.33	0.66
	Elastic	1.0	0.37	0.37
	Plastic	1.0	0.33	0.33
	Elastic	0.40	0.26	0.65
	Plastic	0.50	0.33	0.66

TABLE 24. TRANSFORMATION FACTORS FOR ONE WAY ELEMENTS [9] (cont.)

Edge conditions and loading diagrams	Range of behaviour	Load factor K_L	Mass factor K_M	Load mass factor K_{LM}
	Elastic	1.0	0.24	0.24
	Plastic	1.0	0.33	0.33
	Elastic	0.87	0.52	0.60
	Plastic	1.0	0.56	0.56

$$\frac{I_R^2}{2\frac{K_M}{K_L}M_R} = \frac{R_{Ru}X_E}{2} + R_{Ru}\left(X_m - X_E\right) \tag{106}$$

where $I_R = \frac{1}{2}F_R t_d$.

Consequently, the balance between kinetic energy and strain energy can be stated in terms of the actual total mass M_R, the actual total peak force F_R and the actual resistance R_u, as far as the factor K_M/K_L is introduced. This factor is the load mass factor K_{LM} that is also given in Table 24.

In the case of blast loads, the total force F_R can normally be assumed to be uniformly distributed over the length or the surface of the structural element. For one way elements, $F_R = f_R L$, where L is the total length. When the mass is also uniformly distributed, $M_R = m_R L$; the balance equation can also be written in terms of the specific real quantities f_R, m_R and r_{Ru} (force, mass and resistance per unit length or per unit area, respectively):

$$\frac{i_R^2}{2K_{LM}m_R} = \frac{r_{Ru}X_E}{2} + r_{Ru}\left(X_m - X_E\right) \tag{107}$$

where $i_R = \frac{1}{2}f_R t_d$ is the real specific impulse.

166

When the load and the mass are uniformly distributed, integration of the SDOF equation of motion can be done in terms of specific quantities (i.e. $F = K_{L}f_{R}$ and $M = K_{M}m_{R}$ in Fig. 66), as far as the stiffness of the spring is defined as the relationship between the maximum deflection X and the specific force f_{R}. Stiffness values in Table 25 are given in this way.

When specific quantities are used, ultimate resistance is also introduced as specific resistance r_{Ru}. Values of resistance R_{Ru} and specific resistance r_{Ru} are given in Table 26 for some typical one way elements as a function of the plastic bending moments of the cross-sections: M_P (tensile forces at lower face) and M_N (tensile forces at upper face).

III.1.1. Design for flexure

The design objective is to provide adequate flexural strength and ductility so that the kinetic energy delivered by the impulsive load is absorbed by the strain energy developed by the member in defecting to X_m. To maintain structural integrity and functionality, the deformation of the structural element needs to meet the acceptable criteria (e.g. support rotation $\theta < 4°$) as given in Table 29.

The following steps may be considered in the design process (adapted from Refs [9, 44]):

(a) The resistance–deflection function (R–X in Fig. 66) is defined. For one way structural elements, R_{Ru} or r_{Ru} can be taken from Table 26, once the plastic moment is known.

For computing the plastic moment, three types of reinforced concrete cross-section are considered.

Type I: the concrete is effective in resisting the moment. The concrete cover over the reinforcement on both surfaces of the structural element remains intact. In this case, for a rectangular cross-section with no compression reinforcement, the ultimate resisting moment M_{pl} can be computed as (see Eq. (48)):

$$M_{pl} = A_{s}f_{ds}\left(d - \frac{a}{2}\right)$$
(108)

where

A_{s} is the area of steel in tension;
f_{ds} is the dynamic design stress of the reinforcing steel;

TABLE 25. ELASTIC, ELASTO-PLASTIC AND EQUIVALENT ELASTIC STIFFNESSES FOR ONE WAY ELEMENTS [9]

Edge conditions and loading diagrams	Elastic stiffness K_e	Elasto-plastic stiffness K_{ep}	Equivalent elastic stiffness K_E
	$\dfrac{384E \cdot I}{5L^4}$	—	$\dfrac{384E \cdot I}{5L^4}$
	$\dfrac{48E \cdot I}{L^3}$	—	$\dfrac{48E \cdot I}{L^3}$
	$\dfrac{185E \cdot I}{L^4}$	$\dfrac{384E \cdot I}{5L^4}$	$\dfrac{160E \cdot I^*}{L^4}$
	$\dfrac{107E \cdot I}{L^3}$	$\dfrac{48E \cdot I}{L^3}$	$\dfrac{106E \cdot I^*}{L^3}$
	$\dfrac{384E \cdot I}{L^4}$	$\dfrac{384E \cdot I}{5L^4}$	$\dfrac{307E \cdot I^*}{L^4}$
	$\dfrac{192E \cdot I}{L^3}$	$\dfrac{48E \cdot I^\dagger}{L^3}$	$\dfrac{192E \cdot I^*}{L^3}$
	$\dfrac{8E \cdot I}{L^4}$	—	$\dfrac{8E \cdot I}{L^4}$

Edge conditions and loading diagrams	Elastic stiffness	Elasto-plastic stiffness	Equivalent elastic stiffness
	K_e	K_{ep}	K_E
	$\dfrac{3E \cdot I}{L^3}$	—	$\dfrac{3E \cdot I}{L^3}$
	$\dfrac{56.4E \cdot I}{L^3}$	—	$\dfrac{56.4E \cdot I}{L^3}$

Note: —: data not available; *: valid only if $MN = Mp$; †: valid only if $MN < Mp$; E: Young's modulus; I: flexural inertia (see Figs 67 and 68).

d is the distance from the extreme compression fibre to the centroid of the tension reinforcement;

and a is the depth of the equivalent rectangular stress block in the concrete, given by

$$a = \frac{A_s f_{ds}}{0.85b \cdot f_{dc}}$$

where

b is the width of the compression face;

and f_{dc} is the dynamic ultimate compressive strength of the concrete.

To avoid brittle compression failures, the reinforcement ratio should be kept under 75% of the value that would produce simultaneous failure of concrete and steel at ultimate resisting moment conditions. When there is compression reinforcement A'_s, the ultimate moment is:

TABLE 26. ULTIMATE RESISTANCES FOR ONE WAY ELEMENTS [9]

Edge conditions and loading diagrams	Ultimate resistance
	$r_u = \dfrac{8M_p}{L^2}$
	$R_u = \dfrac{4M_p}{L}$
	$r_u = \dfrac{4\left(M_N + 2M_p\right)}{L^2}$
	$R_u = \dfrac{2\left(M_N + 2M_p\right)}{L}$
	$r_u = \dfrac{8\left(M_N + M_p\right)}{L^2}$
	$R_u = \dfrac{4\left(M_N + M_p\right)}{L}$
	$r_u = \dfrac{2M_N}{L^2}$
	$R_u = \dfrac{M_N}{L}$
	$R_u = \dfrac{6M_p}{L}$

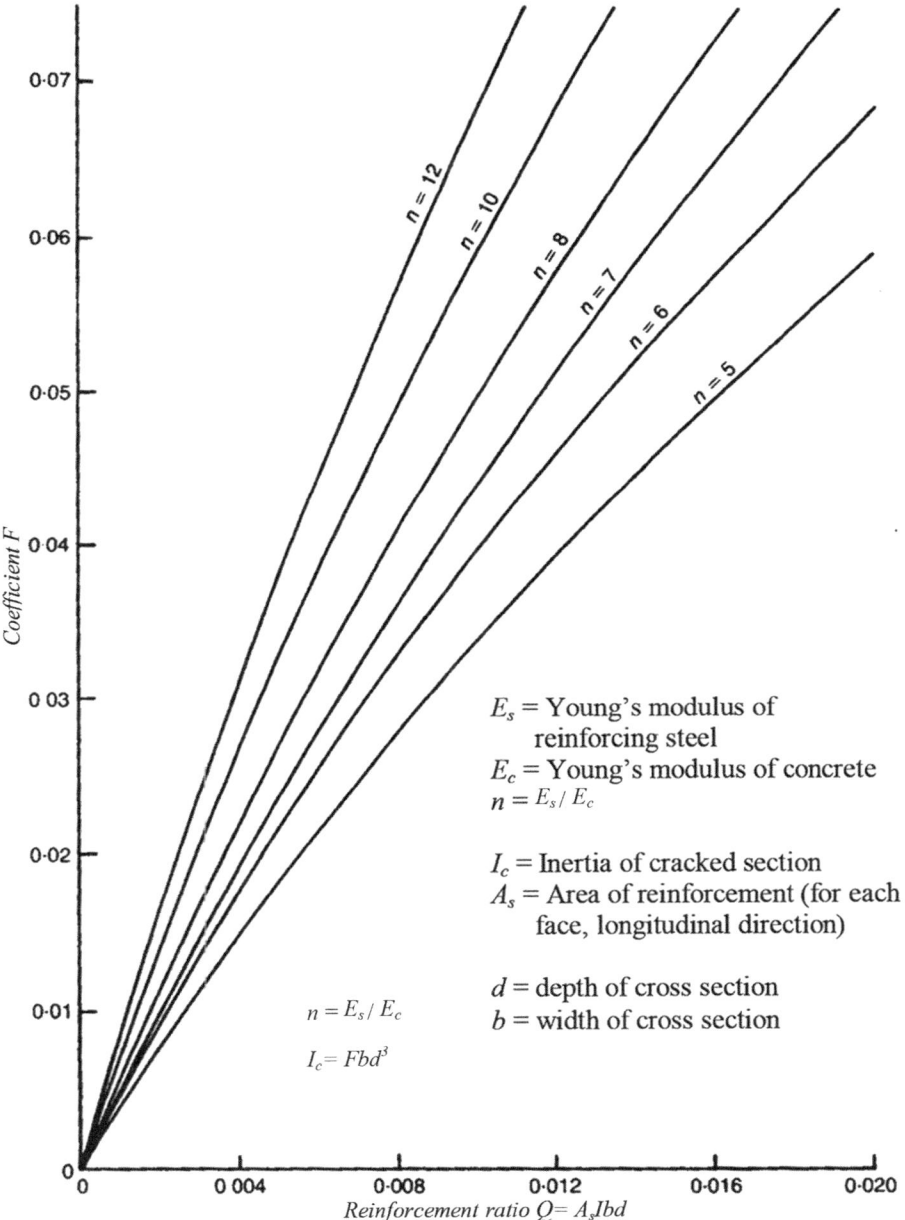

FIG. 67. *Coefficient for moment of inertia of cracked sections with tension reinforcement only (adapted from Ref. [9]).*

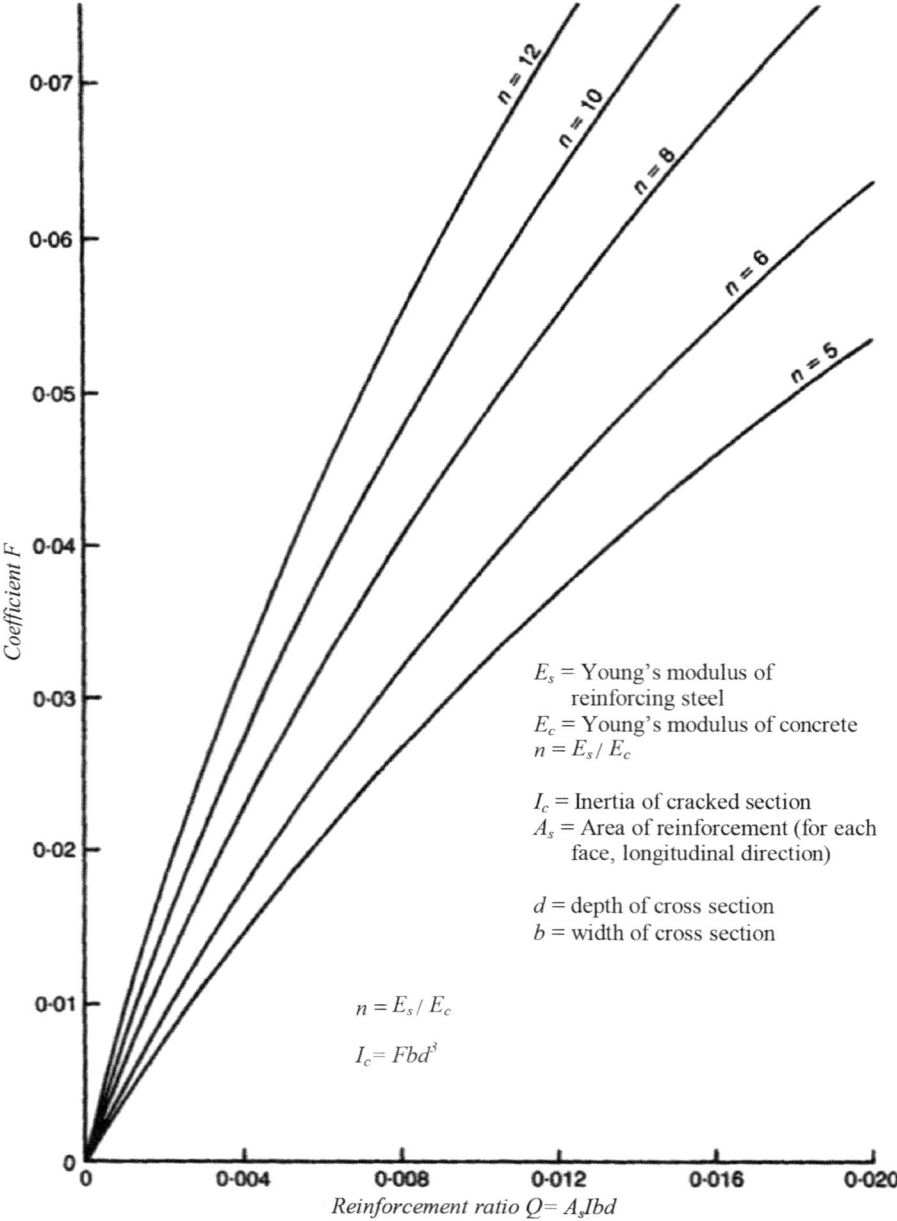

FIG. 68. *Coefficient for moment of inertia of cracked sections with equal reinforcement on opposite faces (adapted from Ref. [9]).*

172

$$M_{\text{pl}} = \left(A_{\text{s}} - A'_{\text{s}}\right)f_{\text{ds}}\left(d - \frac{a}{2}\right) + A'_{\text{s}}f_{\text{ds}}\left(d - d'\right) \qquad (109)$$

where

d' is the distance from the extreme compression fibre to the centroid of the compression reinforcement;

and a is the depth of the equivalent rectangular stress block in the concrete, given by

$$a = \frac{\left(A_{\text{s}} - A'_{\text{s}}\right)f_{\text{ds}}}{0.85b \cdot f_{\text{dc}}} \quad .$$

To avoid brittle compression failures, the difference between the tension reinforcement ratio and the compression reinforcing ratio should be kept under 75% of the value that would produce simultaneous failure of concrete and steel at ultimate resisting moment conditions. On the other hand, Eq. (109) is valid only when compression steel reaches a value of f_{ds} at the ultimate resisting moment. When this condition is not met, Eq. (108) is to be used.

Type II: the concrete is crushed and not effective in resisting the moment. Compression reinforcement is required to resist the moment. The concrete cover over the reinforcement on both surfaces of the structural element remains intact. In this case, for a rectangular cross-section, the ultimate resisting moment M_{pl} can be computed by:

$$M_{\text{pl}} = A_{\text{s}}f_{\text{ds}}d_{\text{c}} \qquad (110)$$

where

A_{s} is the area of tension or compression reinforcement, whichever is less;

and d_{c} is the distance between the centroids of the compression and tension reinforcement.

Type III: the concrete cover over the reinforcement on both surfaces of the structural element is completely disengaged. Equal tension and compression reinforcement, properly tied together, are required to resist the moment. In this case, Eq. (110) can be used.

In the equations above, the dynamic design stresses of the reinforcing steel f_{ds} and those of the concrete f_{dc} are obtained from Table 27 as a function of the type of stress. On the other hand, Table 28 gives the dynamic increase factors, with respect to static values, which can be credited for blast loadings.

To complete the definition of the resistance–deflection function, the first portion of the function is defined by the equivalent elastic stiffness K_E. Table 25 gives the elastic stiffness for one way elements. The resistance, R_{Ru} or r_{Ru}, divided by the elastic stiffness, yields the elastic limit response X_E.

(b) A first estimate of the response time t_m is computed and it is confirmed that it is long when compared with the pulse time t_d (e.g. $t_m \geq 3t_d$). If this condition is not met, the load cannot be considered impulsive and the procedure is not appropriate (the procedure in Section III.2 is then used). An estimate of the response time t_m can be obtained as:

$$t_m \approx \frac{I_R}{R_{Ru}} \quad \text{or} \quad t_m \approx \frac{i_R}{r_{Ru}} \tag{111}$$

(c) The load mass factor K_{LM} is determined from Table 24, and either the total mass M_R or the specific mass m_R, depending on how the balance equation has been formulated (i.e. total or specific quantities).

(d) Equation (106) or Eq. (107) is solved for X_m and compared with allowable values of deformation (e.g. Table 29). If deformation exceeds the allowable value, the design should be modified and one should then go to step (a).

TABLE 27. DYNAMIC DESIGN STRESSES FOR REINFORCED CONCRETE

Type of stress	Dynamic design stress	
	Concrete f_{dc}	Reinforcing bars f_{ds}
Bending	f_{dcu}	f_{dy} [a] $f_{dy} + (f_{du} - f_{dy})/4$ [b]
Shear	f_{dcu}	f_{dy}
Compression	f_{dcu}	f_{dy}

[a] Protection category 1 is intended for the protection of personnel and equipment through attenuation of blast pressures and to shield them from the effects of primary and secondary fragments or falling portions of the structure.

[b] Protection category 2 is intended for the prevention of the collapse of structural elements.

TABLE 28. DYNAMIC DESIGN STRESSES FOR REINFORCED CONCRETE

Type of stress	Concrete f_{dcu}/f_{cu}	Reinforcing bars f_{dy}/f_y	Reinforcing bars f_{du}/f_u	Structural steel f_{dy}/f_y [a]	Structural steel f_{du}/f_u
Bending	1.25	1.20	1.05	1.20	1.05
Shear	1.00	1.10	1.00	1.20	1.05
Compression	1.15	1.10	—[b]	1.10	—[b]

[a] Steel with a minimum specified f_y of 500 MPa or less may be enhanced by the average strength increase factor of 1.10.
[b] — : data not available.

TABLE 29. DEFORMATION LIMITS FOR STRUCTURAL ELEMENTS IN BENDING [44]

	Protection category[a] 1 — Allowable support rotation θ (°)	1 — Allowable ductility ratio $\mu = X_m/X_E$	2 — Allowable support rotation θ (°)	2 — Allowable ductility ratio $\mu = X_m/X_E$
Reinforced concrete beam and slabs[b]	2	n.a.[c]	4	n.a.[c]
Structural steel beams and plates[d]	2	10	12	20

[a] Protection category 1 is intended for the protection of personnel and equipment through attenuation of blast pressures and to shield them from the effects of primary and secondary fragments or falling portions of the structure. Protection category 2 is intended for the prevention of the collapse of structural elements.
[b] Shear reinforcement in the form of open or closed links should be provided in slabs for $\theta > 1°$. Closed links should be provided in all beams.
[c] n.a.: not applicable.
[d] Adequate bracing should be provided to ensure the given levels of ductile behaviour.

III.1.2. Design for shear

After the flexural design, the required quantity of shear reinforcement is to be determined. Shear failure needs to be avoided, since it is a brittle failure. Maximum shear is developed when the bending resistance reaches the ultimate value, either R_{Ru} or r_{Ru}. Consequently, the shear reinforcement is to be obtained from the bending resistance of the structural element, not from the applied load.

Typically, in blast resistant design, a distinction is made between 'diagonal shear' and 'direct shear'. 'Diagonal shear' is the shear associated with the flexural response of an element. It includes 'diagonal tension', 'diagonal compression' and 'punching shear'. It is the kind of shear commonly taken into account by reinforced concrete design codes (e.g. Refs [42, 46, 118]), resulting either from beam (one way) bending or from the punching (two way) action of plates or shells around localized loads or supports.

On the other hand, 'direct shear' failure is characterized by the rapid propagation of a transverse crack (i.e. vertical crack when the structural element is horizontal) located at the supports where the maximum shear stresses occur. This phenomenon is associated with the nearly instantaneous reaction force in response to the blast. It normally appears in near field blasts. Failure of this type is possible even in members reinforced for diagonal shear. It should be noted that stirrups that are placed perpendicular to the plane of a wall or slab provide no resistance to direct shear, since the failure plane is also perpendicular. In practice, diagonal reinforcement is introduced to prevent direct shear failure when the design support rotation exceeds 2° (unless the slab is simply supported) or when the section is in tension (as in containment cells). In the rest of the cases, direct shear is resisted by the concrete. Diagonal reinforcement consists of inclined bars which extend from the support into the slab element.

For diagonal shear, shear reinforcement is obtained using the national reinforced concrete design codes (e.g. Refs [42, 46, 118]) for the shear forces corresponding to the ultimate bending resistance, either R_{Ru} or r_{Ru}.

For direct shear at the supports, the capacity V_{ds} can be computed from:

$$V_{ds} = V_c + A_{sd} f_{ds} \sin\alpha \tag{112}$$

where

A_{sd} is the total area of the diagonal bars at the support;
f_{ds} is the dynamic design stress of the reinforcing steel;
α is the angle formed by the plane of the diagonal reinforcement and the longitudinal reinforcement;

and $V_c = 0.18 f_{dc} \cdot b \cdot d$ for $\theta \le 2°$ or simple supports, and $V_c = 0$ for $\theta > 2°$ or a section in tension.

Parameter b is the width of the cross-section and d is the distance from the extreme compression fibre to the centroid of the tension reinforcement.

As for the diagonal tension shear, shear reinforcement is to be obtained using the shear forces corresponding to the ultimate bending resistance, either R_{Ru} or r_{Ru}. The shear force at the face of the support is used.

III.2. RESPONSE TO QUASI-STATIC/DYNAMIC LOADING

In this case, the blast load can still be idealized as a triangular pressure–time function with zero rise time or as having other idealized pulse shapes for which response charts based on SDOF analyses are available. In the quasi-static/dynamic response regimes, the duration t_d of the applied load is longer compared to the response time t_m of the element.

III.2.1. Design for flexure

The design process follows a workflow which is similar to the one described above for the impulse load. The steps may be as follows (adapted from Ref. [44]):

(a) The resistance–deflection function (R–X in Fig. 66) should be defined. This step is the same as for the impulse load.
(b) The natural period T of the system should be calculated:

$$T = 2\pi \sqrt{\frac{M \cdot X_E}{R_u}} = 2\pi \sqrt{\frac{K_M M_R X_E}{K_L R_{Ru}}}$$

$$= 2\pi \sqrt{\frac{K_{LM} M_R X_E}{R_{Ru}}} = 2\pi \sqrt{\frac{K_{LM} m_R X_E}{r_{Ru}}}$$

(113)

(c) The appropriate SDOF response chart should be referred to (Fig. 69) for an elastic-plastic system under idealized load to obtain the ductility demand, $\mu = X_m/X_E$, and, hence, X_m and associated rotation θ. This should be compared with the allowable values of deformation (e.g. Table 29). If deformation exceeds the allowable value, the design should be modified and one should go to step (1).
(d) The appropriate SDOF response chart should be referred to (Fig. 70) for an elastic-plastic system under idealized load to obtain the response time t_m. It

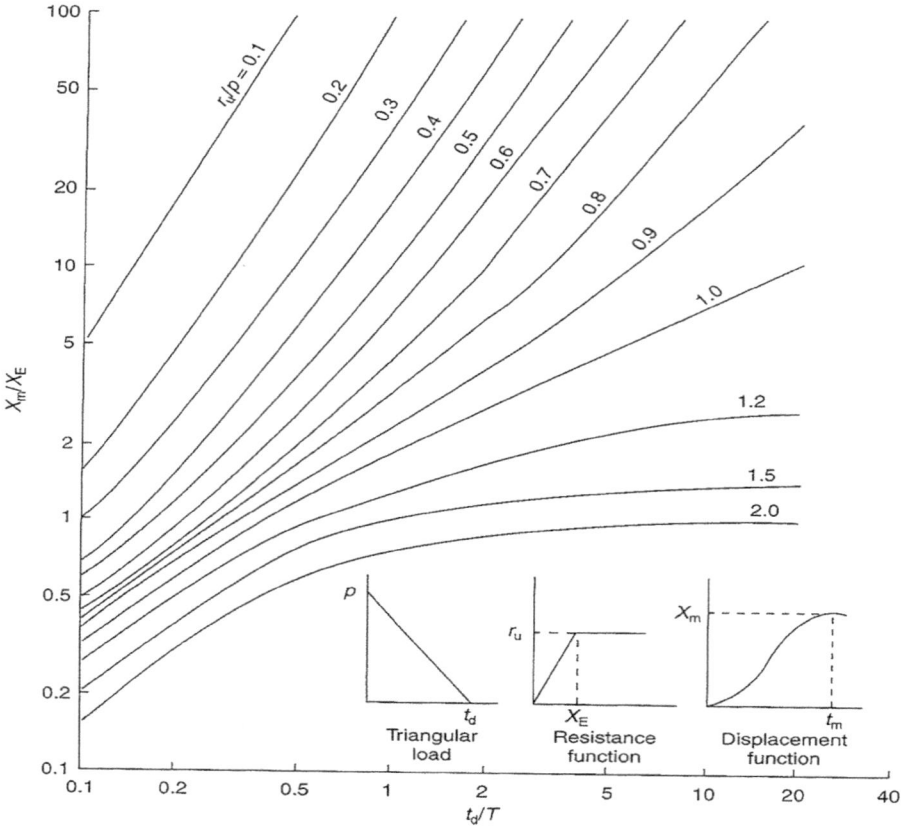

FIG. 69. Maximum deflection of the elastic-plastic single degree of freedom system for a triangular load [9].

should be confirmed that the response time t_m is short when compared with the pulse time t_d (e.g. $t_m < 3t_d$). If this condition is not met, then the load can be considered impulsive and the present procedure is not appropriate (one should then go to the procedure in Section III.1).

III.2.2. Design for shear

After completing the flexural design of the element, the required quantity of shear reinforcement is to be determined, following the same procedure as in the impulse load case.

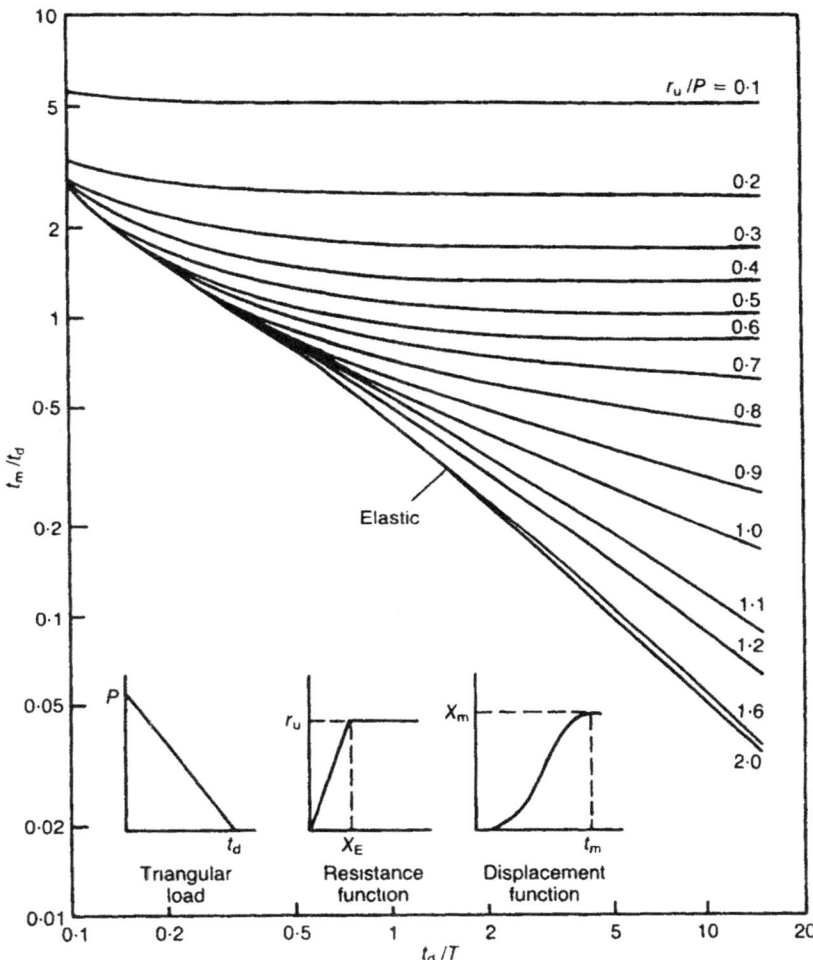

FIG. 70. Maximum response time of the elastic-plastic single degree of freedom system for a triangular load [9].

III.2.3. Dynamic reactions

When a reinforced concrete element is loaded dynamically, the loads transferred to the supports are known as the dynamic reactions. The magnitude of these reactions is a function of both the total resistance R_{Ru} and the total load F_R applied to the element; both of them vary with time. The dynamic reaction V may be expressed in a general form as a linear combination $V = \alpha R_{Ru} + \beta F_R$. The coefficients α and β can be obtained from Ref. [44].

III.3. DETAILING

The design procedures and rules given above assume that reinforced concrete detailing allows for the achievement of the required ductility. Detailing rules can be found in Ref. [9] and a good summary is contained in Ref. [119]. The basic rules are as follows:

— The specified compressive strength of the concrete is not to be less than 20 MPa.
— Use of high strength concrete (i.e. specified compressive strength more than about 35 MPa) is not recommended, since it could experience more brittle failure modes when subjected to inelastic yielding.
— Concrete sections with tension reinforcement only are not permitted, since the structural member needs to resist load reversals and rebound. Compression reinforcement is to be at least one half the required tension reinforcement.
— The minimum flexural reinforcing ratio in a slab or a wall is 0.18% in each face and each direction.
— The absence of transverse shear reinforcement can be justified for sections with no ductility demand, which basically remain in the elastic range (e.g. type I cross-sections). Shear reinforcement is always to be used for near field blasts (see Section 2.3.1 for a definition of 'near field').
— Except for cases in which the absence can be justified, the minimum transverse shear reinforcement ratio is 0.15% and the maximum spacing of stirrups is limited to $d/2$ in type I sections or $d_c/2$ in type II or III sections.
— Transverse shear reinforcement is obtained for the critical section and this quantity of reinforcement is uniformly distributed throughout the structural element.

III.4. EXAMPLES

Two examples are presented in this section. The first example corresponds to the design of a cantilever wall to resist blast loading; the second example deals with the design of a reinforced concrete wall panel against dynamic loading.

III.4.1. Design example 1: Reinforced concrete cantilever to resist impulsive loading

A reinforced concrete cantilever wall is to be designed to withstand the impulse load due to 120 kg of TNT detonated at ground level with a stand-off distance of 5 m, for protection category 2 (collapse prevention). The height of the wall is 3.5 m and it is symmetrically reinforced with steel, having a yield stress of 460 MPa, an ultimate stress of 550 MPa and a reinforcement ratio $\rho_s = 0.4\%$ (each face, vertical direction). The design compressive strength of the concrete is 40 MPa with a density of 2400 kg/m^3.

III.4.1.1. Determination of blast load parameters

The ground explosion can be assumed to be a hemispherical (or surface) blast, and the relevant blast loading factors at the wall can be obtained from Fig. 4. It should be noted that Fig. 4 corresponds to spherical charges and, therefore, for a hemispherical ground blast, the TNT equivalent mass needs to be multiplied by 1.8 before entering the curves. Hence:

$$W = 1.8 \times 120 \ (\text{kg}) = 216 \ (\text{kg})$$

$$Z = \frac{5 \ (\text{m})}{\sqrt[3]{1.8 \times 120 \ (\text{kg})}} = 0.83 \frac{(\text{m})}{(\text{kg})^{1/3}} \tag{114}$$

From Fig. 4:

Specific impulse: $\dfrac{i_s}{\sqrt[3]{W}} \approx 183 \dfrac{(\text{Pa} \cdot \text{s})}{\sqrt[3]{(\text{kg})}} \quad \rightarrow \quad i_s = 1.1 \ (\text{MPa} \cdot \text{ms})$ \hfill (115)

Peak overpressure: $P_{so} \approx 1.37 \ (\text{MPa})$

However, for the design of the wall, the reflected pressure is needed, since it is the pressure which will be sustained by the wall. The reflected peak overpressure P_{ro} can be obtained using Eq. (9):

$$P_{ro} = P_{so} \ (4 \lg_{10} P_{so} + 0.5) \quad \rightarrow \quad P_{ro} = 13.7 \ (4 \lg_{10} 13.7 + 0.5)$$
$$= 69.14 \ (\text{bar} \approx 6.9 \, \text{MPa})$$

The reflected specific impulse can also be obtained from Fig. 4:

$$\frac{i_r}{\sqrt[3]{W}} \approx 724 \frac{(\text{Pa} \cdot \text{s})}{\sqrt[3]{(\text{kg})}} \quad \rightarrow \quad i_r = 4.4 \ (\text{MPa} \cdot \text{ms}) \tag{116}$$

Equivalent triangular pulse duration: $t_d = \dfrac{2 i_r}{P_{ro}} = \dfrac{2 \times 4.4}{6.9} = 1.28 \ (\text{ms})$

III.4.1.2. Design for flexure

For a unit width of cantilever retaining wall of height H and effective depth d_c:

(a) The resistance–deflection function (R–X in Fig. 66) should be defined. Specific resistance r_{Ru} is taken from Table 26 for a cantilever beam:

$$r_{Ru} = \frac{2M_N}{H^2}$$

where M_N is the plastic moment. If a type II cross-section is assumed, as in Eq. (108):

$$M_N = M_{pl} = A_s f_{ds} d_c$$

where

A_s is the area of steel in tension;

and f_{ds} is the dynamic design stress of the reinforcing steel.

From Table 27, if protection category 2 is assigned to the wall:

$$f_{ds} = f_{dy} + \frac{(f_{du} - f_{dy})}{4} \tag{117}$$

Table 28 gives a dynamic increase factor of 1.20 for the yield stress and 1.05 for the ultimate tensile stress:

$$f_{dy} = 1.20 \times 460 = 552 \ (\text{MPa})$$

$$f_{du} = 1.05 \times 550 = 578 \ (\text{MPa})$$

$$f_{ds} = 552 + \frac{578 - 552}{4} = 559 \ (\text{MPa}) \tag{118}$$

From the reinforcing ratio:

$$A_s = 0.004 b \cdot d_c$$

Therefore:

$$r_{Ru} = \frac{2M_N}{H^2} = \frac{2A_s f_{ds} d_c}{H^2} = \frac{2 \times 0.004b \cdot d_c \cdot f_{ds} \cdot d_c}{H^2}$$

$$= \frac{0.008 f_{ds} \cdot b \cdot d_c^2}{H^2} = 365 \, d_c^2 \; (\text{kPa})$$
(119)

where d_c is in metres.

The first portion of the resistance–deflection function is defined by the equivalent elastic stiffness K_E. Table 25 gives the elastic stiffness for a cantilever beam:

$$K_E = \frac{8E \cdot I}{H^4} = \frac{8 \times 28 \times 10^9}{3.5^4} \, 0.0196b \cdot d_c^3 = 29.2 d_c^3 \; (\text{MPa/m})$$
(120)

where d_c is in metres.

A Young's modulus of 28 GPa has been assumed. Figure 68 gives a coefficient of 0.0196 for a reinforcing ratio of 0.4% and $n = 200/28 = 7$. Consequently, the elastic deflection X_E is:

$$X_E = \frac{r_{Ru}}{K_E} = \frac{365 d_c^2}{29\,200 d_c^3} = \frac{12.5}{d_c} \; (\text{mm})$$
(121)

where d_c is in metres.

(b)   An estimate of the response time t_m can be obtained from Eq. (111):

$$t_m = \frac{i_r}{r_{Ru}} = \frac{4.4 \, (\text{MPa} \cdot \text{ms})}{0.365 \, (\text{MPa}) \, d_c^2} = \frac{12.05 \, (\text{ms})}{d_c^2}$$

where d_c is in metres.

(c)   The load mass factor K_{LM} from Table 24 is 0.66, and the specific mass m_R is:

$$m_R = 2400 \; (\text{kg/m}^3) \, d_c$$
(122)

where d_c is in metres.

(d) After substitution, balance in Eq. (107) gives, for a maximum acceptable rotation of 4° at the base (protection category 2 in Table 29):

$$\frac{4.4^2}{2 \times 0.66 \times 2400 d_c} = \frac{0.365 d_c^2 \times 0.0125}{2 d_c} + 0.365 d_c^2 \left(3.5 \tan 4° - \frac{0.0125}{2 d_c} \right)$$

which results in the following equation in d_c:

$$d_c^3 - 0.00255 d_c^2 - 0.0684 = 0 \quad \rightarrow \quad d_c = 0.41 \text{ (m)}$$

The response time t_m is then $12.05/d_c^2 = 72$ ms, which is long when compared with the equivalent pulse time $t_d = 1.28$ ms (i.e. $t_m \geq 3t_d$). Hence, the load can be considered impulsive and the procedure is appropriate. From the assumed reinforcing steel ratio of 0.4%:

$$A_s = 0.004 \, b \cdot d_c = 0.004 \times 100 \text{ (cm)} \times 41 \text{(cm)}$$
$$= 16.40 \text{ (cm}^2/\text{m)} \quad \rightarrow \quad \varnothing 20 \text{mm bars at 150 mm centres}$$

Therefore, Ø20 mm bars are used at 15 cm centres on each face, and the overall section thickness, with 40 mm cover, is given by:

$$T_c = 40 + 10 + 410 + 10 + 40 \text{ (mm)} = 51 \text{ (cm)}$$

Normally, in reinforced concrete design, the total thickness will be adjusted to the nearest multiple of 5 cm.

III.4.1.3. Design for shear

Diagonal shear

Commonly used reinforced concrete design codes (e.g. Refs [42, 46, 118]) require verification of shear resulting from bending and of punching shear in plates or shells around localized loads or supports. In the present case, only one way bending shear is present. One way bending shear capacity is usually checked at a distance of about one effective depth from the supports. The wall structure is isostatic; hence, the shear demand at a distance d_c from the base is given by:

$$V_u = r_{Ru}(H - d_c) = 0.365 d_c^2 (H - d_c) = 365 \times 0.41^2 \times (3.5 - 0.41) = 190 \text{ (kN/m)}$$

It should be noted that the shear demand is computed from the bending specific capacity r_{Ru}, not from the actual load on the wall.

Following, for instance, Eurocode 2 [46], shear capacity for members with shear reinforcement perpendicular to longitudinal reinforcement is the smaller of:

$$V_{d1} = \frac{A_{sw}}{s} d_c f_{ds} \cot\theta \qquad 1 \le \cot\theta \le 2.5 \qquad \text{(diagonal tension)}$$

and

$$V_{d2} = \frac{1}{(\cot\theta + \tan\theta)} b_w d_c \nu_1 f_{cs} \qquad 1 \le \cot\theta \le 2.5 \qquad \text{(diagonal compression)}$$

where

A_{sw} is the cross-sectional area of the shear reinforcement;
s is the spacing of the stirrups;
b_w is the minimum width between tension and compression chords;

and ν_1 is the strength reduction factor for concrete cracked in shear; a value of 0.50 is taken.

Tables 27 and 28 give, for shear:

$$f_{dy} = 1.10 \times 460 = 506 \, (\text{MPa})$$

$$f_{ds} = f_{dy} = 506 \, (\text{MPa})$$

$$f_{dc} = f_{dcu} = f_{cu} = 40 \, (\text{MPa})$$

which gives, taking $\cot\theta = 1.0$:

$$V_{d1} = \frac{A_{sw}}{s} 0.41 \times 506 \times 1.0 = 207 \frac{A_{sw}}{s} \, (\text{MN/m})$$

where A_{sw}/s is in metres

and $V_{d2} = \frac{1}{(1.0 + 1.0)} 1.0 \times 0.41 \times 0.50 \times 40 = 4100 \, (\text{kN/m})$

Hence, the required amount of reinforcement is:

$$V_{d1} = V_u \quad \rightarrow \quad 207 \frac{A_{sw}}{s} = 0.190 \quad \rightarrow \quad \frac{A_{sw}}{s}$$
$$= 9.18 \times 10^{-4} \, (\mathrm{m^2/m}) = 9.18 \, (\mathrm{cm^2/m})$$

If spacing s is taken as 20 cm ($\approx d_c/2$), A_{sw} = 1.84 cm^2 at 20 cm centres for each 1 m wide strip of wall. For design purposes, Ø8 mm stirrups are used at 15 cm (horizontal) × 20 cm (vertical) centres, to comply with the minimum ratio of 0.15%.

Direct shear

Since the acceptable rotation angle at the base of the wall is larger than 2°, additional reinforcement is introduced at the connection with the support to resist direct shear. Direct shear is checked at the face of the support; in this case, at the base of the wall. The shear demand at the base is given by:

$$V_u = r_{Ru} H = 0.365 d_c^2 H = 365 \times 0.41^2 \times 3.5 = 215 \, (\mathrm{kN/m})$$

It should be noted again that the shear demand is computed from the bending specific capacity r_{Ru}, not from the actual load on the wall.

Then, if diagonal bars at a 45° angle are used, Eq. (112) gives the required amount of reinforcement as:

$$A_{sd} = \frac{V_u}{f_{ds} \sin \alpha} = \frac{0.215}{506 \times \sin 45°} = 6.0 \, (\mathrm{cm^2})$$

for each 1 m wide strip of wall.

Therefore, a row of Ø12 mm diagonal bars at 15 cm (horizontal) centres are introduced into the design.

The final reinforcement arrangement is shown in Fig. 71.

III.4.2. Design example 2: Reinforced concrete wall panel to resist dynamic loading

This example shows the design of a fixed ended reinforced concrete wall panel for protection category 1 against a specific blast loading. The panel is 500 mm thick and spans vertically over an effective height of 6 m. As in the previous example, it is symmetrically reinforced with steel, having a yield stress of 460 MPa, an ultimate stress of 550 MPa and reinforcement ratios $\rho_s = \rho'_s = 0.4\%$ (each face, vertical direction), with effective depths $d = 450$ mm

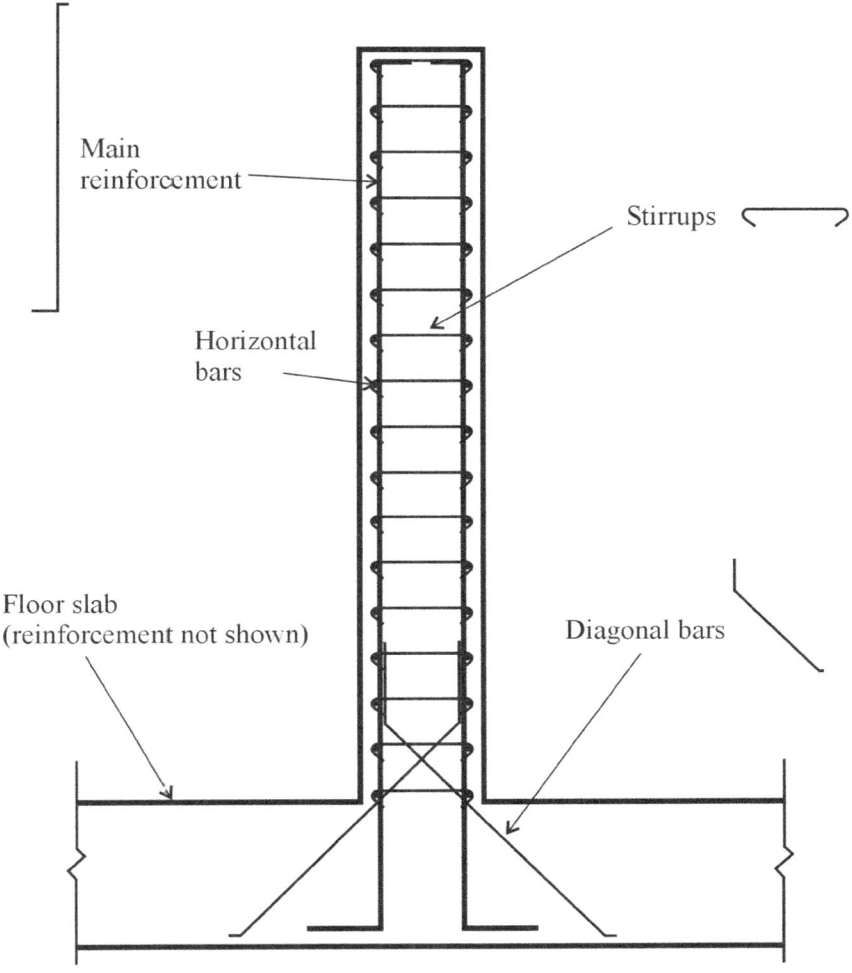

FIG. 71. Design example 1: arrangement of reinforcement.

and $d' = 50$ mm. The design compressive strength of the concrete is 40 MPa with a density of 2400 kg/m³.

III.4.2.1. Blast load parameters

For this example, the specified blast loading threat is idealized as a triangular pressure–time function with a peak reflected pressure $P_{ro} = 200$ kPa and a time duration $t_d = 50$ ms.

III.4.2.2. Design for flexure

For a unit width of wall having an effective height H, an effective depth d and for a Type 1 section, the design steps are as follows:

(a) The resistance–deflection function (R–X in Fig. 66) should be defined. Specific resistance r_{Ru} is taken from Table 26 for a simply supported beam:

$$r_{Ru} = \frac{8(M_P + M_N)}{H^2}$$

where M_p is the plastic moment. If a type I cross-section is assumed, the contribution from the reinforcement in the compression zone can normally be ignored, as in Eq. (110):

$$M_P = M_{pl} = A_s f_{ds}\left(d - \frac{a}{2}\right)$$

with:

$$a = \frac{A_s f_{ds}}{0.85b \cdot f_{dc}}$$

From Tables 27 and 28, if protection category 1 is assigned to the wall:

$$f_{dy} = 1.20 \times 460 = 552 \text{ (MPa)}$$

$$f_{ds} = f_{dy} = 552 \text{ (MPa)}$$

$$f_{dcu} = 1.25 \times 40 = 50 \text{ (MPa)}$$

$$f_{dc} = f_{dcu} = f_{cu} = 50 \text{ (MPa)}$$

From the reinforcing ratio:

$$A_s = A_s' = 0.004b \cdot d = 0.004 \times 1.0 \times 0.45 = 18 \text{ (cm}^2/\text{m)}$$

Therefore:

$$a = \frac{A_s f_{ds}}{0.85b \cdot f_{dc}} = \frac{18 \times 10^{-4} \times 552 \times 10^6}{0.85 \times 1.0 \times 50 \times 10^6} = 2.34 \text{ (cm)}$$

$$M_{pl} = A_s f_{ds} \left(d - \frac{a}{2} \right) = 18 \times 10^{-4} \times 552 \times (0.450 - 0.012)$$

$$= 435 \, (\text{kN} \cdot \text{m} \cdot \text{m}^{-1})$$

$$r_{Ru} = \frac{8(M_N + M_P)}{H^2} = \frac{8 \times (435 + 435)}{6^2} = 193.3 \, (\text{kPa})$$

The first portion of the resistance–deflection function is defined by the equivalent elastic stiffness K_E. Table 25 gives the elastic stiffness for a fixed ended (clamped) beam:

$$K_E = \frac{307 E \cdot I}{H^4} = \frac{307 \times 28 \times 10^9}{6^4} 0.0196 b \cdot d^3 = 11.85 \, (\text{MPa/m})$$

A Young's modulus of 28 GPa has been assumed. Figure 68 gives a coefficient of 0.0196 for a reinforcing ratio of 0.4% and $n = 200/28 = 7$. Consequently, the elastic deflection X_E is:

$$X_E = \frac{r_{Ru}}{K_E} = \frac{0.1933}{11.85} = 16.3 \, (\text{mm})$$

(b) The natural period T of the system is calculated from Eq. (113):

$$T = 2\pi \sqrt{\frac{K_{LM} m_R X_E}{r_{Ru}}} = 2\pi \sqrt{\frac{0.66 \times 1200 \times 0.0163}{193300}} = 51.4 \, (\text{ms})$$

where the load mass factor K_{LM} has been taken from Table 24 and the specific mass m_R is $2400 \times 0.50 = 1200 \, \text{kg/m}^2$.

(c) The ductility demand $\mu = X_m/X_E$ is obtained from Fig. 69 for $r_{Ru}/P_{ro} = 193.3/200 = 0.97$ and $t_d/T = 50/51.4 = 0.97$. A value of $\mu \approx 1.8$ is read from the figure. Then, the maximum displacement is found to be $X_m = 1.8 \times 16.3 = 29.34 \, \text{mm}$.

For protection category 1, the allowable rotation at plastic hinges is 2° (Table 29). However, if no stirrups are to be used, then the rotation needs to be further reduced to 1°. Since the rotation at the mid-height plastic hinge is twice the rotation at the supports, the maximum allowable deflection at mid-height will be:

$$X_{lim} = \frac{H}{2} \tan 0.5° = \frac{6000 \, (\text{mm})}{2} 0.00873 = 26.2 \, (\text{mm}) \leq X_m = 29.3 \, (\text{mm})$$

Hence, the maximum deformation X_m exceeds the limit deformation X_{lim} for a panel without stirrups. Stirrups will need to be added, but the design for flexure is valid.

(d) From Fig. 70, the normalized response time $t_m/t_d = 0.58$ is obtained for $r_{Ru}/P_{ro} = 193.3/200 = 0.97$ and $t_d/T = 50/51.4 = 0.97$. The response time is less than the pulse duration (i.e. $t_m \ll 3t_d$) and, therefore, the procedure is appropriate.

III.4.2.3. Design for shear

Diagonal shear

As in the previous example, in the present case, only one way bending shear is present. One way bending shear capacity is usually checked at a distance of about one effective depth from the supports. The shear demand at a distance d from the supports is given by:

$$V_u = r_{Ru} \left(\frac{H}{2} - d\right) = 193.3\,(3.0 - 0.45) = 493\ (\text{kN/m})$$

It should be noted that the shear demand is computed from the bending specific capacity r_{Ru}, not from the actual load on the wall.

Following Eurocode 2 [46], shear capacity for members with shear reinforcement perpendicular to longitudinal reinforcement is the smaller of:

$$V_{d1} = \frac{A_{sw}}{s}\,0.9d \cdot f_{ds}\cot\theta \qquad 1 \leq \cot\theta \leq 2.5 \qquad \text{(diagonal tension)}$$

and

$$V_{d2} = \frac{1}{(\cot\theta + \tan\theta)}\,b_w\,0.9d \cdot v_1 f_{cs} \qquad 1 \leq \cot\theta \leq 2.5 \text{ (diagonal compression)}$$

where

A_{sw} is the cross-sectional area of the shear reinforcement;
s is the spacing of the stirrups;
b_w is the minimum width between tension and compression chords;

and v_1 is the strength reduction factor for concrete cracked in shear; a value of 0.50 is taken.

Tables 27 and 28 give, for shear:

$$f_{dy} = 1.10 \times 460 = 506 \,(\text{MPa})$$

$$f_{ds} = f_{dy} = 506 \,(\text{MPa})$$

$$f_{dc} = f_{dcu} = f_{cu} = 40 \,(\text{MPa})$$

which gives, taking $\cot \theta = 2.0$:

$$V_{d1} = \frac{A_{sw}}{s}\, 0.9 \times 0.45 \times 506 \times 2.0 = 410 \frac{A_{sw}}{s}\ (\text{MN/m})$$

and

$$V_{d2} = \frac{1}{(2.0 + 0.5)}\, 1.0 \times 0.9 \times 0.45 \times 0.50 \times 40 = 3240 \,(\text{kN/m})$$

Hence, the required amount of reinforcement is:

$$V_{d1} = V_u \quad \rightarrow \quad 410 \frac{A_{sw}}{s} = 0.493 \quad \rightarrow \quad \frac{A_{sw}}{s}$$

$$= 12.0 \times 10^{-4} \,(\text{m}^2/\text{m}) = 12.0 \ (\text{cm}^2/\text{m})$$

If spacing s is taken as 20 cm ($\approx d_c/2$), $A_{sw} = 2.40$ cm^2 at 20 cm centres for each 1 m wide strip of panel. For design purposes, Ø8 mm stirrups are used at 20 cm (horizontal) × 20 cm (vertical) centres.

Direct shear

Since the design is for a section type I, with maximum rotations far less than 2°, no additional reinforcement is introduced at the connection with the supports to resist direct shear.

REFERENCES

[1] INTERNATIONAL ATOMIC ENERGY AGENCY, External Human Induced Events in Site Evaluation for Nuclear Power Plants, IAEA Safety Standards Series No. NS-G-3.1, IAEA, Vienna (2002).

[2] INTERNATIONAL ATOMIC ENERGY AGENCY, External Events Excluding Earthquakes in the Design of Nuclear Power Plants, IAEA Safety Standards Series No. NS-G-1.5, IAEA, Vienna (2003).

[3] INTERNATIONAL ATOMIC ENERGY AGENCY, Protection against Internal Fires and Explosions in the Design of Nuclear Power Plants, IAEA Safety Standards Series No. NS-G-1.7, IAEA, Vienna (2004).

[4] INTERNATIONAL ATOMIC ENERGY AGENCY, Safety Aspects of Nuclear Power Plants in Human Induced External Events: General Considerations, Safety Reports Series No. 86, IAEA, Vienna (2017).

[5] INTERNATIONAL ATOMIC ENERGY AGENCY, Safety Aspects of Nuclear Power Plants in Human Induced External Events: Margin Assessment, Safety Reports Series No. 88, IAEA, Vienna (2017).

[6] KOECHLIN, P., POTAPOV, S., Classification of soft and hard impacts — Application to aircraft crash, Nucl. Eng. Des. **239** (2009) 613–618.

[7] RIERA, J.D., On the stress analysis of structures subjected to aircraft impact forces, Nucl. Eng. Des. **8** (1968) 415–426.

[8] NUCLEAR ENERGY INSTITUTE, Methodology for Performing Aircraft Impact Assessment for New Plant Designs, Rev. 8P, Rep. prepared by ERIN Engineering & Research, Washington, DC (2011).

[9] JOINT DEPARTMENTS OF THE ARMY, NAVY, AND AIR FORCE, USA, Structures to Resist the Effects of Accidental Explosions, Rev. 1, Rep. TM 5-1300/NAVFAC P-397/AFR 88-22 (revised as United Facilities (UFC), Structures to Resist the Effects of Accidental Explosions, Rep. UFC-3-340-02 (2008)), Washington, DC (1990).

[10] BAKER, W.E., et al. (Eds), Explosion Hazards and Evalutation, Fundamental Studies in Engineering Series, Vol. 5, Elsevier, Amsterdam (1983).

[11] PINGLI, Y., Explosionseinwirkungen auf Stahlbetonplatten, Institut für Massivbau und Baustofftechnologie, Vol. 11, Karlsruhe Univ. (1991).

[12] KINNEY, G.F., GRAHAM, K.J., Explosive Shocks in Air, 2nd edn, Springer, Berlin (1985).

[13] NEWMARK, N.M., "External blast", Proc. Int. Conf. on the Planning and Design of Tall Buildings, Vol. 1b, Lehigh University (1972) 661–676.

[14] NEDERLANDSE ORGANISATIE VOOR TOEGEPAST NATUURWETENSCHAP-PELIJK ONDERZOEK (TNO), Methods for the Determination of Possible Damage to People and Objects Resulting from Release of Hazardous Materials, The Green Book, Rep. CPR 16E, Voorburg, Netherlands (1992) Ch. 2.

[15] AMERICAN SOCIETY OF CIVIL ENGINEERS, Design of Blast-resistant Buildings in Petrochemical Facilities, 2nd edn, Task Committee on Blast Resistant Design of the Petrochemical Committee of the Energy Division, ASCE, Reston, VA (2010).

[16] GLASSTONE, S., DOLAN, P.J., The Effects of Nuclear Weapons, 3rd edn, US Department of Defense and US Department of Energy, Washington, DC (1977) 127–141.

[17] PANDEY, A.K., KUMAR, R., PAUL, D.K., TRIKHA, D.N., Non-linear response of reinforced concrete containment structure under blast loading, Nucl. Eng. Des. **236** (2006) 993–1002.

[18] REMENNIKOV, A.M., ROSE, T.M., Modelling Blast Loads on Buildings in Complex City Geometries, Faculty of Engineering, Papers, Wollongong Univ., (2005).

[19] JOHANSSON, M., LARSEN, O.P., LAINE, L., "Experiments and analyses of explosion at an intersection", Proc. 20th Symp. on Military Aspects of Blast and Shock, Oslo (2008).

[20] AMERICAN INSTITUTE OF CHEMICAL ENGINEERS, Guidelines for Vapor Cloud Explosions, Pressure Vessel Burst, BLEVE and Flash Fires, 2nd edn, Wiley, New York (2010).

[21] BJERKETVEDT, D., BAKKE, J., van WINGERDEN, K., Gas Explosion Handbook, GexCon AS, Bergen (1992).

[22] DI NENNO, P.J., et al. (Eds), SFPE Handbook of Fire Protection Engineering, 4th edn, National Fire Protection Association, Quincy, MA (2008).

[23] HIETANIEMI, J., MIKKOLA, E., Design Fires for Fire Safety Engineering, VTT Working Papers 139, VTT Technical Research Centre of Finland, Espoo (2010).

[24] LEE, B.T., Heat Release Rate Characteristics of Some Combustible Fuel Sources in Nuclear Power Plants, NBSIR 85-3195, National Bureau of Standards, Gaithersburg, MD (1985).

[25] NUCLEAR REGULATORY COMMISSION, EPRI/NRC-RES Fire PRA Methodology for Nuclear Power Facilities — Final Report, Rep. NUREG/CR-6850, EPRI 1011989, Washington, DC (2005).

[26] HAUPTABTEILUNG FÜR DIE SICHERHEIT DER KERNANLAGEN, Stellungnahme der HSK zur Sicherheit der schweizerischen Kernkraftwerke bei einem vorsätzlichen Flugzeugabsturz, Rep. HSK-AN-4626, Swiss Federal Nuclear Safety Inspectorate, Würenlingen (2003).

[27] ABBASI, T., ABBASI, S.A., The boiling liquid expanding vapour explosion (BLEVE): Mechanism, consequence assessment, management, J. Hazard. Mater. **141** (2007) 489–519.

[28] DOROFEEV, S.B., et al., Fireballs from deflagration and detonation of heterogeneous fuel-rich clouds, Fire Saf. J. **25** (1995) 323–336.

[29] NEDERLANDSE ORGANISATIE VOOR TOEGEPAST NATUURWETENSCHAP-PELIJK ONDERZOEK (TNO), Methods for the Calculation of Physical Effects due to Releases of Hazardous Materials (Liquids and Gases), 3rd edn, The Yellow Book, Rep. CPR 14E, The Hague (1997) Ch. 6.

[30] BABRAUSKAS, V., Estimating large pool fire burning rates, Fire Technol. **19** (1983) 251–261.

[31] REW, P.J., HULBERT, W.G., DEAVES, D.M., Modelling of thermal radiation from external hydrocarbon pool fires, Process Saf. Environ. Prot. **75** (1997) 81–89.

[32] BEYLER, C.L., "Fire hazards calculations for large, open hydrocarbon fires", SFPE Handbook of Fire Protection Engineering, 4th edn, National Fire Protection Association, Quincy, MA (2008) Ch. 10, Section 3.

[33] SIKANEN, T., HOSTIKKA, S., SILDE, A., "Experimental and numerical simulations of liquid spreading and fires after aircraft impact", Proc. 21st Int. Conf. on Structural Mechanics in Reactor Technology (SMiRT 21), Pre-conference Seminar on Fire Safety in Nuclear Power Plant Installations, Cologne (2011).

[34] LUTHER, W., MÜLLER, W.C., FDS simulation of the fuel fireball from a hypothetical commercial airliner crash on a generic nuclear power plant, Nucl. Eng. Des. **239** (2009) 2056–2069.

[35] McGRATTAN, K.B., MILES, S., "Modeling enclosure fires using computational fluid dynamics (CFD)", SFPE Handbook of Fire Protection Engineering, 4th edn, National Fire Protection Association, Quincy, MA (2008) Ch. 8, Section 3.

[36] SUGANO, T., et al., Full-scale aircraft impact test for evaluation of impact force, Nucl. Eng. Des. **140** (1993) 373–385.

[37] SILDE, A., HOSTIKKA, S., KANKKUNEN, A., Experimental and numerical studies of liquid dispersal from a soft projectile impacting a wall, Nucl. Eng. Des. **241** (2011) 617–624.

[38] VEPSÄ, A., AATOLA, S., CALONIUS, K., HAKOLA, I., HALONEN, M., "Impact 2014 (IMPACT2014) and structural mechanics analysis of soft and hard impacts (SMASH)", SAFIR2014 — The Finnish Research Programme on Nuclear Power Plant Safety 2011–2014, VTT Technology Series **213** (2015) 555–591.

[39] SCHMEHL, R., MAIER, G., WITTIG, S., "CFD analysis of fuel atomization, secondary droplet breakup and spray dispersion in the premix duct of a LPP combustor", Proc. 8th Int. Conf. on Liquid Atomization and Spray Systems, Pasadena, CA, 2000.

[40] NUCLEAR REGULATORY COMMISSION, Fire Dynamics Tools (FDTs) — Quantitative Fire Hazard Analysis Methods for the U.S. Nuclear Regulatory Commission Fire Protection Inspection Program, Rep. NUREG-1805, Office of Standards Development, Washington, DC (2004).

[41] DRITTLER K., GRUNER, P., Zur Auslegung kerntechnischer Anlagen gegen Einwirkungen von außen. Teilaspekt. Berechnung von Kraft-Zeit-Verläufen beim Aufprall deformierbarer Flugkörper auf eine starre Wand, Tech. Rep. IRS-W-14, Institut für Reaktorsicherheit der Technischen Überwachungs-Vereine e.V., Cologne (1975).

[42] AMERICAN CONCRETE INSTITUTE, Code Requirements for Nuclear Safety-Related Concrete Structures (ACI 349-06) and Commentary, ACI, Farmington Hills, MI (2007).

[43] BISCHOFF, P.H., PERRY, S.H., Compressive behaviour of concrete at high strain rates, Mater. Struct. **24** (1991) 425–450.

[44] CORMIE, D., MAYS, G., SMITH, P. (Eds), Blast Effects on Buildings, 2nd edn, Thomas Telford, London (2009).

[45] SCHULER, H., MAYRHOFER, C., THOMA, K., "Experimental determinations of damage parameter and implementation into a new damage law", Proc. 11th Int. Symp. on Interaction of the Effects of Munitions with Structures, Mannheim (taken from SCHULER, H., Experimentelle und numerische Untersuchungen zur Schädigung von stoßbeanspruchtem Beton, PhD Thesis, Fraunhofer Institut für Kurzzeitdynamik, Ernst-Mach-Institut, Freiburg (2004)) (2003).

[46] EUROPEAN COMMITTEE FOR STANDARDIZATION, Eurocode 2 — Design of Concrete Structures, Rep. EN 1992-1-2:2004, Brussels (2004).

[47] JONES, N., Structural Impact, Cambridge University Press, Cambridge (1989).

[48] EUROPEAN COMMITTEE FOR STANDARDIZATION, Eurocode 3 — Design of Steel Structures, Rep. EN 1993-1-2:2005 and Erratum EN 1993-1-2:2005/AC:2009, Brussels (2009).

[49] LEE, J., FENVES, G.L., Plastic-damage model for cyclic loading of concrete structures, J. Eng. Mech. **124** (1998) 892–900.

[50] KRUTZIK, N.J., "Reduction of the dynamic response by aircraft crash on building structures", Proc. 9th Int. Conf. on Structural Mechanics in Reactor Technology (SMiRT 9), Lausanne (1987).

[51] ORBOVIC, N., ELGOHARY, M., LEE, N., BLAHOIANU, A., "Tests on reinforced concrete slabs with pre-stressing and with transverse reinforcement under impact loading", Proc. 20th Int. Conf. on Structural Mechanics in Reactor Technology (SMiRT 20), Paper 2015, Espoo, Finland (2009).

[52] ORBOVIC, N., BLAHOIANU, A., "Tests to determine the influence of transverse reinforcement on perforation resistance of RC slabs under hard missile impact", Proc. 22nd Int. Conf. on Structural Mechanics in Reactor Technology (SMiRT 22), San Francisco, CA (2013).

[53] CHADMAIL, J.F., KRUTZIK, N.J., "Equivalent loading due to airplane impact taking into account the non-linearities of impacted reinforced concrete buildings", Proc. 7th Int. Conf. on Structural Mechanics in Reactor Technology (SMiRT 7), Paper J/3, Chicago, IL (1983).

[54] COMITÉ EURO-INTERNATIONAL DU BÉTON, CEB-FIP Model Code 1990, Redwood Books, Trowbridge, Wiltshire (1990).

[55] RAMBACH, J.M., TARALLO, F., "Simple analytical models for beams and slabs under soft impacts at medium speed", Proc. 16th Int. Conf. on Nuclear Engineering ICONE16-48583, Orlando, FL (2008).

[56] SAARENHEIMO, A., et al., Experimental and numerical studies on projectile impacts, Rakenteiden Mekaniikka **42** (2009) 1–37.

[57] LI, Q.M., REID, S.R., WEN, H.M., TELFORD, A.R., Local impact effects of hard missiles on concrete targets, Int. J. Impact Eng. **32** (2005) 224–284.

[58] OECD, NUCLEAR ENERGY AGENCY, Improving Robustness Assessment Methodologies for Structures Impacted by Missiles (IRIS_2012), Rep. NEA/CSNI/R(2014) 5, OECD Publishing, Paris (2014).

[59] STEVENSON, J.D., Structural damping values as a function of dynamic response stress and deformation levels, Nucl. Eng. Des. **60** (1980) 211–237.

[60] EIBL, J., KRUTZIK, N.J., "Applicability limits of finite element models for simulation of shock transfer processes in concrete structures", Proc. 17th Int. Conf. on Structural Mechanics in Reactor Technology (SMiRT 17), Prague (2003).

[61] KRUTZIK, N.J., Applicability Limits of Finite Element Models for Simulation of Shock Transfer Processes in Nuclear Power Plant Structures due to Impact Loading, PhD Thesis, Karlsruhe Univ. (2002).

[62] AMERICAN SOCIETY OF CIVIL ENGINEERS, Seismic Analysis of Safety-Related Nuclear Structures, Rep. ASCE 4-98, ACI, Reston, VA (1998).

[63] DEUTSCHES INSTITUT FÜR NORMUNG, Bauteile aus Stahl- und Spannbeton in kerntechnischen Anlagen, Rep. DIN 25449, DIN, Berlin (2006).

[64] KERNTECHNISCHER AUSSCHUSS (German Nuclear Safety Standards Commission), Design of Nuclear Power Plants against Seismic Events–Part 3: Civil Structures, Rep. KTA 2201.3 (2013-11), Salzgitter (2013).

[65] ARROS, J., DOUMBALSKI, N., Analysis of aircraft impact to concrete structures, Nucl. Eng. Des. **237** (2007) 1241–1249.

[66] LIVERMORE SOFTWARE, LS-DYNA Theory Manual, Livermore, CA (2006).

[67] ASSOCIATION FRANÇAISE POUR LES RÈGLES DE CONCEPTION DE CONSTRUCTION ET DE SURVEILLANCE EN EXPLOITATION DES MATÉRIELS DES CHAUDIÈRES ELECTRO NUCLÉAIRES, EPR Technical Code for Civil Works, Rep. AFCEN ETC-C:2010, Paris–La Defense (2010).

[68] NUCLEAR REGULATORY COMMISSION, Development of Floor Design Response Spectra for Seismic Design of Floor-Supported Equipment or Components, Regulatory Guide 1.122, Rev. 1, US NRC, Washington, DC (1978).

[69] KRUTZIK, N.J., TROPP, R., Verification of the local structural response of building structures in the anchorage areas of heavy components, Nucl. Eng. Des. **141** (1993) 385–393.

[70] CHOPRA, A., Dynamics of Structures: Theory and Applications to Earthquake Engineering, Prentice Hall, Upper Saddle River, NJ (1995).

[71] KRUTZIK, N.J., "Simplified design of components and systems against aircraft crash induced loads using verified response spectra", in Proc. 7th Int. Conf. on Structural Mechanics in Reactor Technology (SMiRT 7), Paper J/3a, Chicago (1983).

[72] HENKEL, F.-O., "Broadening/smoothening of floor response spectra — how to deal with 'needle peaks'", Proc. 21st Int. Conf. on Structural Mechanics in Reactor Technology (SMiRT 21), Div-V, Paper 1332011, New Delhi (2011).

[73] KERNTECHNISCHER AUSSCHUSS (German Nuclear Safety Standards Commission), Bauwerksabdichtung von Kernkraftwerken, Rep. KTA 2501 (2010-11), Salzgitter (2010).

[74] INTERNATIONAL ORGANIZATION FOR STANDARDIZATION, Fire-Resistance Tests — Elements of Building Construction — Part 1: General Requirements, Rep. ISO 834-1, ISO, Geneva (1999).

[75] ASTM INTERNATIONAL, Standard Test Methods for Fire Tests of Building Construction and Materials, Rep. ASTM E119-14, ASTM, West Conshohocken, PA (2014).

[76] INTERNATIONAL ATOMIC ENERGY AGENCY, Engineering Safety Aspects of the Protection of Nuclear Power Plants against Sabotage, IAEA Nuclear Security Series No. 4, IAEA, Vienna (2007).

[77] CANADIAN NUCLEAR SAFETY COMMISSION, Design of Reactor Facilities: Nuclear Power Plants, Rep. REGDOC-2.5.2, CNSC, Ottawa (2014).

[78] AMERICAN INSTITUTE OF STEEL CONSTRUCTION, Specification for Safety-Related Steel Structures for Nuclear Facilities, Rep. AISC N690-06, AISC, Chicago (2006).

[79] DET NORSKE VERITAS, Design Against Accidental Loads, Recommended Practice, Rep. DNV-RP-C204 (2010).

[80] KENNEDY, R.P., A review of procedures for the analysis and design of concrete structures to resist missile impact effects, Nucl. Eng. Des. **37** (1976) 183–203.

[81] UNITED STATES DEPARTMENT OF ENERGY, Accident Analysis for Aircraft Crash into Hazardous Facilities, Rep. DOE-STD-3014-96, USDOE, Washington, DC (1996).

[82] JOWETT, J., KINSELLA, K., "Soft missile perforation analysis of small and large scale concrete slabs", Structures under Shock and Impact (BULSON, P.S., Ed.), Elsevier (1989) 121–132.

[83] DEUTSCHES INSTITUT FÜR NORMUNG, Lay-out of Nuclear Installations for Aircraft D Crash Loading, Rep. DIN Special Report NKE FB3 Nr.1- 93, DIN, Berlin (1993).

[84] LI, Q.M., "Impact effects on concrete", Advances in Protective Structures Research (HAO, H., LI, Z., Eds), CRC Press (2012) Ch. 10.

[85] CHANG, W.S., Impact of solid missiles on concrete barriers, J. Struct. Div. Am. Soc. Civ. Eng. **107** (1981) 257–271.

[86] HOSSAIN, Q.A., et al., Structures, Systems, and Components Evaluation Technical Support Document for the DOE Standard, Accident Analysis for Aircraft Crash into Hazardous Facilities, Technical Rep. UCRL-ID-123577, Lawrence Livermore Natl Lab., Livermore, CA (1996).

[87] TELAND, J.A., A Review of Empirical Equations for Missile Impact Effects on Concrete, FFI/RAPPORT-97/05856, Norwegian Defence Research Establishment, Kjeller (1998).

[88] KOJIMA, I., An experimental study on local behavior of reinforced concrete slabs to missile impact, Nucl. Eng. Des. **130** (1991) 121–132.

[89] BARR, P., Guidelines for the Design and Assessment of Concrete Structures Subjected to Impact, AEA Technology (1990).

[90] ADELI, H., AMIN, A.M., Local effects of impactors on concrete structures, Nucl. Eng. Des. **88** (1985) 301–317.

[91] BERRIAUD, C., SOKOLOVSKY, A., GUERAUD, R., DULAC, J., LABROT, R., Local behaviour of reinforced concrete walls under missile impact, Nucl. Eng. Des. **45** (1978) 457–469.

[92] FULLARD, K., BARR, P., Development of design guidance for low velocity impacts on concrete floors, Nucl. Eng. Des. **115** (1989) 113–120.

[93] FULLARD, K., BAUM, M.R., BARR, P., The assessment of impact on nuclear power plant structures in the United Kingdom, Nucl. Eng. Des. **130** (1991) 113–120.

[94] DEGEN, P.P., Perforation of reinforced concrete slabs by rigid missiles, J. Struct. Div. Am. Soc. Civ. Eng. **106** (1980) 1623–1642.

[95] TUOMALA, M., CALONIUS, K., SAARENHEIMO, A., VÄLIKANGAS, P., "Numerical studies on pre-stressed impact loaded concrete walls", Proc. 20th Int. Conf. on Structural Mechanics in Reactor Technology (SMiRT 20), Paper 1921, Espoo, Finland (2009).

[96] KAR, A.K., Residual velocity for projectiles, Nucl. Eng. Des. **53** (1979) 87–95.

[97] McVAY, M.K., Spall Damage of Concrete Structures, Rep. SL-88-22, Department of the Army, US Army Corps of Engineers, Washington, DC (1988).

[98] MILLS, C.A., "The design of concrete structures to resist explosions and weapon effects", Proc. 1st Int. Conf. on Concrete for Hazard Protection, Edinburgh (1987) 61–73.

[99] BASLER, E., Lokale Schadenwirkungen an Betonplatten durch Sprengladungen, Rep. B3113.10-2, E. Basler & Partner, Zurich (1982).

[100] ANDERSSON, P., VAN HEES, P., Performance of cables subjected to elevated temperatures, Fire Saf. Sci. **9** (2005) 1121–1132.

[101] TAYLOR, G., et al., Electrical Cable Test Results and Analysis During Fire Exposure (ELECTRA-FIRE) — A Consolidation of Three Major Fire-induced Circuit and Cable Failure Experiments Performed Between 2001 and 2011, Rep. NUREG-2128, NRC, Washington, DC (2013).

[102] GAY, L., GRACIA, R., WIZENNE, E., Thermal malfunction criteria of fire safety electrical equipment in nuclear power plants, Fire Mater. **37** (2013) 151–159.

[103] TANAKA, T.J., NOWLEN, S.P., KORSAH, K., WOOD, R.T., ANTONESCU, C.E., Impact of smoke exposure on digital instrumentation and control, Nucl. Technol. **143** (2003) 152–160.

[104] KESSLER, G., et al., The Risks of Nuclear Energy Technology — Safety Concepts of Light Water Reactors, Springer, Berlin (2014).

[105] DRITTLER, K., GRUNER, P., The force resulting from impact of fast-flying military aircraft upon a rigid wall, Nucl. Eng. Des. **37** (1976) 245–248.

[106] BUNDESMINISTERIUM UMWELT, RSK-Leitlinien für Druckwasser Reaktoren, 3rd edn, Cologne (1981).

[107] MUTO, K., et al., "Full-scale aircraft impact test for evaluation of impact force. Part 2: Analysis of the results", Proc. 10th Int. Conf. on Structural Mechanics in Reactor Technology (SMiRT 10), Anaheim, CA (1989).

[108] JACKSON, P. (Ed.), IHS Jane's All the World's Aircraft 2013-2014, IHS Global, Coulsdon, United Kingdom (2013).

[109] WOLF, J.P., BUCHER, K.M., SKRIKERUD, P.E., Response of equipment to aircraft impact, Nucl. Eng. Des. **47** (1978) 169–193.

[110] KIRKPATRICK, S., et al., "Evaluation of aircraft impact analysis methodologies for nuclear safety applications", Proc. 22nd Int. Conf. on Structural Mechanics in Reactor Technology (SMiRT 22), San Francisco, CA (2013).

[111] MLAKAR, P.E., et al., American Society of Civil Engineers, Pentagon Building Performance Report, Reston, VA (2003).

[112] CHAI, S.T., MASON, W.H., Landing Gear Integration in Aircraft Conceptual Design, Virginia Polytechnic Institute and State Univ., Blacksburg, VA (1996).

[113] POLITECNICO DI MILANO, Aircraft Systems — Lecture Notes, Dipartimento di Ingegneria Aerospaziale, Milan (2004).

[114] HENKEL, F.-O., KLEIN, D., "Variants of analysis of the load case airplane crash", Proc. of 19th Int. Conf. on Structural Mechanics in Reactor Technology (SMiRT 19), Paper J03/2, Toronto (2007).

[115] HENKEL, F.-O., WÖLFEL, H., Building concepts against airplane crash, Nucl. Eng. Des. **79** (1984) 397–409.

[116] BIGGS, J.M., Introduction to Structural Dynamics, McGraw-Hill, New York (1964).

[117] SMITH, P.D., HETHERINGTON, J.G., Blast and Ballistic Loading of Structures, Butterworth-Heinemann, Oxford, United Kingdom (1994).

[118] BRITISH STANDARDS INSTITUTION, Structural Use of Concrete — Part 1: Code of Practice for Design and Construction, Rep. BS 8110, BSI, London (1997).

[119] AMERICAN SOCIETY OF CIVIL ENGINEERS, Blast Protection of Buildings, Rep. ASCE/SEI 59-11, ASCE, Reston, VA (2011).

ABBREVIATIONS AND ACRONYMS

BLEVE	boiling liquid expanding vapour explosion
CEA–EDF	French Alternative Energies and Atomic Energy Commission–Électricité de France
CFD	computational fluid dynamics
CRIEPI	Central Research Institute of the Electric Power Industry
DBE	design basis event
DEE	design extension event
DLF	dynamic load factor
ISSC-EBP	International Seismic Safety Centre extra-budgetary programme
MTOW	maximum take-off weight
NDRC	National Defence Research Committee
RLF	rigid loading function
SDOF	single degree of freedom
SI	International System of Units
SPH	smoothed particle hydrodynamics
SSC	structures, systems and components
TDOF	two degrees of freedom
TNT	trinitrotoluene
VLF	verified loading function

CONTRIBUTORS TO DRAFTING AND REVIEW

Altinyollar, A.	International Atomic Energy Agency
Basu, P.	International Atomic Energy Agency
Beltran, F.	International Atomic Energy Agency
Blahoianu, A.	Canadian Nuclear Safety Commission, Canada
Boroumand, P.	International Atomic Energy Agency
Henkel, F.-O.	Wölfel Beratende Ingenieure, Germany
Hostikka, S.	Aalto University, Finland
Iqbal, J.	Pakistan Atomic Energy Commission, Pakistan
Johnson, J.J.	James J. Johnson & Associates, United States of America
Krutzik, N.J.	NJK Consulting, Germany
Lautkaski, R.	VTT Technical Research Centre, Finland
Markovic, D.	Electricité de France, France
Morita, S.	International Atomic Energy Agency
Orbovic, N.	Canadian Nuclear Safety Commission, Canada
Pino, G.	ITER Consult, Italy
Pisharady, A.	Atomic Energy Regulatory Board, India
Poveda, A.	International Atomic Energy Agency
Rangelow, P.	AREVA, Germany
Ravindra, M.K.	M.K. Ravindra Consulting, United States of America
Ricciuti, R.	CANDU Energy, Canada
Saarenheimo, A.	VTT Technical Research Centre, Finland
Samaddar, S.	International Atomic Energy Agency
Tuomala, M.	Tampere University of Technology, Finland

Varpasuo, P.   Fortum Nuclear Services, Finland

Välikangas, P.   Radiation and Nuclear Safety Authority, Finland

Consultants Meetings

Ottawa, Canada: 28–29 March 2011, 10–14 September 2012

Vienna, Austria: 4–7 October 2011, 12–14 November 2012, 17–21 December 2012,

31 January–1 February 2013, 11–15 November 2013

ORDERING LOCALLY

In the following countries, IAEA priced publications may be purchased from the sources listed below or from major local booksellers.

Orders for unpriced publications should be made directly to the IAEA. The contact details are given at the end of this list.

CANADA

Renouf Publishing Co. Ltd
22-1010 Polytek Street, Ottawa, ON K1J 9J1, CANADA
Telephone: +1 613 745 2665 • Fax: +1 643 745 7660
Email: order@renoufbooks.com • Web site: www.renoufbooks.com

Bernan / Rowman & Littlefield
15200 NBN Way, Blue Ridge Summit, PA 17214, USA
Tel: +1 800 462 6420 • Fax: +1 800 338 4550
Email: orders@rowman.ccm Web site: www.rowman.com/bernan

CZECH REPUBLIC

Suweco CZ, s.r.o.
Sestupná 153/11, 162 00 Prague 6, CZECH REPUBLIC
Telephone: +420 242 459 205 • Fax: +420 284 821 646
Email: nakup@suweco.cz • Web site: www.suweco.cz

FRANCE

Form-Edit
5 rue Janssen, PO Box 25, 75921 Paris CEDEX, FRANCE
Telephone: +33 1 42 01 49 49 • Fax: +33 1 42 01 90 90
Email: formedit@formedit.fr • Web site: www.form-edit.com

GERMANY

Goethe Buchhandlung Teubig GmbH
Schweitzer Fachinformationen
Willstätterstrasse 15, 40549 Düsseldorf, GERMANY
Telephone: +49 (0) 211 49 874 015 • Fax: +49 (0) 211 49 874 28
Email: kundenbetreuung.goethe@schweitzer-online.de • Web site: www.goethebuch.de

INDIA

Allied Publishers
1st Floor, Dubash House, 15, J.N. Heredi Marg, Ballard Estate, Mumbai 400001, INDIA
Telephone: +91 22 4212 6930/31/69 • Fax: +91 22 2261 7928
Email: alliedpl@vsnl.com • Web site: www.alliedpublishers.com

Bookwell
3/79 Nirankari, Delhi 110009, INDIA
Telephone: +91 11 2760 1283/4536
Email: bkwell@nde.vsnl.net.in • Web site: www.bookwellindia.com

ITALY

Libreria Scientifica "AEIOU"
Via Vincenzo Maria Coronelli 6, 20146 Milan, ITALY
Telephone: +39 02 48 95 45 52 • Fax: +39 02 48 95 45 48
Email: info@libreriaaeiou.eu • Web site: www.libreriaaeiou.eu

JAPAN

Maruzen-Yushodo Co., Ltd
10-10 Yotsuyasakamachi, Shinjuku-ku, Tokyo 160-0002, JAPAN
Telephone: +81 3 4335 9312 • Fax: +81 3 4335 9364
Email: bookimport@maruzen.co.jp • Web site: www.maruzen.co.jp

RUSSIAN FEDERATION

Scientific and Engineering Centre for Nuclear and Radiation Safety
107140, Moscow, Malaya Krasnoselskaya st. 2/8, bld. 5, RUSSIAN FEDERATION
Telephone: +7 499 264 00 03 • Fax: +7 499 264 28 59
Email: secnrs@secnrs.ru • Web site: www.secnrs.ru

UNITED STATES OF AMERICA

Bernan / Rowman & Littlefield
15200 NBN Way, Blue Ridge Summit, PA 17214, USA
Tel: +1 800 462 6420 • Fax: +1 800 338 4550
Email: orders@rowman.com • Web site: www.rowman.com/bernan

Renouf Publishing Co. Ltd
812 Proctor Avenue, Ogdensburg, NY 13669-2205, USA
Telephone: +1 888 551 7470 • Fax: +1 888 551 7471
Email: orders@renoufbooks.com • Web site: www.renoufbooks.com

Orders for both priced and unpriced publications may be addressed directly to:
Marketing and Sales Unit
International Atomic Energy Agency
Vienna International Centre, PO Box 100, 1400 Vienna, Austria
Telephone: +43 1 2600 22529 or 22530 • Fax: +43 1 2600 29302 or +43 1 26007 22529
Email: sales.publications@iaea.org • Web site: www.iaea.org/books